Rupture of
the Virtual

John Kim

Rupture of
the Virtual

1
Introduction 1

2
A New Materialism for New Media Studies 19

3
The Origin of the See-Through Graphical Interface 43

4
Encounters with the Material: Krzysztof Wodiczko and Site-Specific Media Art 69

5
Materiality in Site Responsive Media Art 95

6
Rupture of the Virtual 121

7
Conclusion 145

Notes 153
Index 167

Introduction

OPENING

The human sensorium has always been mediated. (Without the "medium" of air or water, the anthropoid ear finds it impossible to hear.) But over the past few decades that condition has greatly intensified. Amplified, shielded, channeled, prosthetized, simulated, stimulated, irritated—our sensorium is more mediated today than ever before. Yet it bothers us less... The microspeaker in the ear, the drug in the blood, the nanosurgical implant, the simulated taste in the mouth—these "enhancements" no longer provoke the apocalyptic excitement they did even a few years ago.[1]

Caroline Jones's observation about the augmentation of human sensory perception echoes an earlier one. Marshall McLuhan's famous claim that the media are extensions of our senses was an early observation about how the media function as a kind of "electronic nervous system" and provide us with post-human capacities to see and hear things not bound by spatial or temporal constraints.[2] Jones' microspeaker in the ear is conceptually linked to McLuhan's earlier claim that the television camera enables us to see things happening on the other side of the globe through discussions of the computer screen as a window to virtual spaces, and films as portals to earlier times and places.

While this claim about the media continues to influence their characterization, this book explores the media's role in producing knowledge about what is immediately under our noses, our physical environment. We are accustomed to thinking about the media as extensions of human perception beyond what we can perceive with unaugmented sensory organs. But to what extent has the origin and construction of the media been guided by those things in our immediate physical surroundings? In this book, I develop a concept of the material in order to investigate the development of modern media and computing devices that enhance knowledge and forms of contact with one's physical surroundings. The term material is used here to distinguish it from the virtual, which has had a lasting influence over ideas concerning the extension and enhancement of perception, like McLuhan's and Jones'. The virtual, however, is a notoriously flexible term, used in a wide variety of contexts.[3] The next chapter specifies a particular definition of the term applied in this book. I use the term broadly and in ways comparable to Anne Friedberg in her book *Virtual Window*, in which she focuses on the "liminally immaterial" image cast by media technologies and their use in the construction of virtual spaces that have little or no corollary to those that exist in one's physical surroundings.[4]

The emphasis on the virtual in media research has occluded insights into the status of the material. A core theoretical argument made in this book is the necessity of rethinking the origin of computing interfaces with an emphasis not on their construction as a virtual window,[5] but on enhancing knowledge of those things in one's immediate physical environment. In light of the proliferation of media that project visual information into the view of one's immediate physical surroundings—a class of computer user interface technologies I refer to as See-Through Graphical Interfaces (STGI)—a notion of the material is needed more than ever. STGIs include a range of devices and phenomena as diverse as Augmented Reality, outdoor projections, site-specific media art, Head-up Displays in aircraft and cars, and others. Recent media research has turned its attention to one's physical environment and its implications for the design and construction of the media and computing devices. By understanding the origin of STGIs, it becomes possible to locate what has been at stake in their development, specifically their enhancement of knowledge and control over one's immediate surroundings. As I argue in Chapter 3 in a history of the STGI, the Head-up Display is a progenitor of modern Augmented Reality-type displays and was originally developed for military fighter aircraft. The Head-up Display pre-dates the invention of many devices integral to the design of the digital computer and its interface.

DEFINING TERMS: THE MEDIUM AND THE IMMEDIATE

The difference between the material and the virtual lies at the intersection of a related linguistic and philosophical distinction, the one between medium and the immediate.[6] Medium derives from the classical Latin, *medius*, for middle, center, midst, intermediate course, intermediary. A medium is a thing, person, or process that links other things, persons, or processes separated by distance, time or inconvenience. A medium has taken on a wide variety of meanings that range across temporal and spatial references both literal and figurative. A medium can be a person who links the past and the present, the living to the dead; and it also refers to the raw material or mode of expression used in an artistic or creative activity, such as the medium of painting. In Raymond Williams' account of the modern usage of the word, he notes that medium has come to stand in for the means of communication that makes possible or facilitates the transmission of a message:

This was particularly clear when the word acquired the first element of its modern sense in the early seventeenth century. Thus to the Sight three things are required, the Object, the Organ and the Medium'. Here a description of the practical activity of seeing, which is a whole and complex process of relationship between the developed organs of sight and the accessible properties of things seen, is characteristically interrupted by the invention of a third term which is given its own properties, in abstraction from the practical relationship. [7]

A medium stands between a message and its receiver. It intervenes in communication, but is also necessary for its transmission. To use a cybernetic formulation, a medium is any intermediate substance through which messages are transmitted from sender to receiver. For Caroline Jones, this intermediate substance can be as basic as air and water, the media for sound, or pigment and canvas as the medium of painting. Raymond Williams focuses on the more modern usage of the term when defining medium as a channel of transmission. For Williams, a medium is any means of communication that augments or extends the conveyance of a message beyond human capabilities, which explains the adoption of the word as applied to the rise of newspapers and other forms of mass communication.[8]

The immediate, on the other hand, derives from the Latin root, *mediare*, to be in the middle, intercede, or to act as an intermediary. Immediate is the negation of the adjectival form of *mediare*, which describes a person or thing in its relation to another as not having a middle or intermediary.[9] Like medium, the immediate also has a wide variety of meanings that range across temporal and spatial references. Immediate can refer to people, such as laymen who have a direct relationship with the gods (without the need of an intermediary, such as a priest); immediate can also refer to intervening time, as in to do something without delay. In the spatial senses of the term it can exhibit two features: unmediated and proximity. By unmediated, the person or thing is perceived directly by the perceiver without the use of an external medium. By proximity, the person or object is within close distance from perceivers such that they can view or hear with their own senses.

What is immediate, however, is more complicated than this definition, because medium and the immediate are often conceived of as mutually determinative; one is defined against the presence or absence of the other. An illuminating example of this is in Denis Diderot's reflections on trompe-l'œil painting in the mid-eighteenth century:

There are several small paintings by Chardin... Nearly all represent fruit and the accessories of a meal. They are nature itself; the objects seem to come forward from the canvas and have a look of reality which deceives the eye... When I look at other artists' paintings I feel I need to make myself a new pair of eyes; to see Chardin's I only need to keep those which nature gave me and use them well... For the porcelain vase is truly porcelain; those olives are really separated from the eye by the water in which they float... Chardin can deceive you and me whenever he wishes. [10]

Diderot's appreciation for Chardin's still lifes lies in their ability to fool a viewer into mistaking the mediated for the immediate. Despite having an immediate relation to the objects (Diderot perceives them with his own eyes), he imagines that the paintings could be mistaken for the things they represent. Chardin's still lifes fool viewers into thinking that they are unmediated, that is, not representations of fruits and the accessories of the meal done in the medium of painting, but the actual objects themselves. Both aspects of the meaning of immediate are evident in this account of Chardin's paintings: Diderot's proximity to the still lifes allows him to perceive them with his own eyes. Nonetheless, he imagines that it is possible to mistake them for the unmediated.

Indeed, Diderot appraises the quality of Chardin's paintings on the basis of their likeness to the unmediated things they represent. In this way the medium and the immediate mutually constitute each other. The paintings Diderot so admires can be understood as a mediated immediacy. Diderot's examination both frames and produces a notion of immediacy through an examination of the medium of painting. This manner of thinking is comparable to Derrida's analysis of the dialectical interplay of speech and writing in *Dissemination*.[11] In this text, Derrida studies Plato's text *Phaedrus*, which concerns Plato's bias towards speech as a privileged means of communication in Greek thought. Writing is understood to be subordinate, because writing was conceived to be in dialectical tension with speech. Writing is mediated, delayed and distanced from the sender in contrast to the immediacy of spoken words.

In the section on the "Pharmakon" in *Dissemination*, Derrida focuses on Plato's definition of the "pharmakon," which Plato uses as a metaphor for writing, but also defines as a drug, recipe, philtre, charm, medicine, spell, artificial colour, and paint. The pharmakon is ambivalent, because it is situated in a flickering and disorienting play of philosophical oppositions:

If the pharmakon is 'ambivalent,' it is because it constitutes the medium in which opposites are opposed, the movement and the play that links them among themselves, reverses them or makes one side cross over into the other (soul/ body, good/ evil, inside/ outside, memory/ forgetfulness, speech/ writing, etc.)....The pharmakon is the movement, the locus, and the play: (the production of) difference... Contradictions and pairs of opposites are lifted from the bottom of this diacritical, differing, deferring, reserve. Already inhabited by *différance*, this reserve, even though it 'precedes' the opposition between different effects, even though it preexists differences as effects, does not have the punctual simplicity of a coincidentia oppositorum. It is from this fund that dialectics draws its reserves. [12]

Speech and writing are discrete forms of communication, but Derrida observes that their meaning is staged in dialectical tension with each other. Speech is that which is inside, linked to memory, immediate and presence, while writing is conceived as its binary opposite: outside, forgetting, mediated and absence. Writing in this sense is secondary, because it is communicated in a medium, and therefore, a less immediate form of communication. But, as Derrida observes, this distinction is artificial, because it is based on a groundless presumption that speech is an unmediated form of communication. Speech, of course, is articulated in the medium of spoken language.

THE MATERIAL: A MEDIATED IMMEDIACY?

The interdependency of the medium and the immediate has influenced the study of the historical development of media technologies and the perceptual logics by which they operate. In their influential book on emerging technologies, *Remediation*, Jay Bolter and Richard Grusin note that the existence of the logic of immediacy in the construction of modern media interfaces in a way that restates Derrida's position on mediation.[13] They argue that the invention of new media has often been guided by an impulse to erase or dissimulate the recognition that it is produced as a medium. In other words, media aspire to a transparency such that viewers do not recognize the mediation. Though Bolter and Grusin make use of the term immediacy, it is meant not as an absence of mediation, but a sensibility that emerges when new media forms refer to older ones. When new media are popularized, older media appear by comparison to be more familiar, intuitive and immediate. A (unintentional or intentional) design strategy is to integrate earlier media forms into new ones in order to produce a sense of immediacy in a user's interactions with it. This integration of earlier forms is what they term remediation. As an example of remediation, Bolter and Grusin consider cut-scenes in video games, non-interactive film clips that are inserted into what is otherwise an interactive experience.[14] Because cut-scenes are an older non-interactive medium, they foreground what is distinctive about the newer. By referencing them, video game play gains a heightened immediacy.

The implication of Bolter and Grusin's argument is that which we understand to be immediate is often the result of the remediation of an earlier form, that is, what we take to be immediate is always already mediated. This is a logic that Derrida's deconstructionist analysis of mediation helps to reveal. Despite the fact that both speech and writing are forms of mediation, speech is attributed a heightened sense of immediacy, because of its difference and presumed anteriority to writing. Bolter and Grusin's position relies on the same underlying conceit: because mediation always already constrains our perception of the world, it is impossible to conceptualize an immediate relation to it. Immediacy only exists in a mediated form, because the immediate is discursively produced by the logic of remediation.

This tendency to distill questions of immediacy to that of another form of mediation pervades media research. It is not only present in Bolter and Grusin's account of remediation, but it is also an assumption in Jones and McLuhan's account of the augmentation of human's sensory abilities. That which we experience through our electronic nervous system has an ontological status indistinguishable from that which we experience through our sensory organs. As I will argue in the next chapter, this has often been rendered to the argument that everything is always already virtual, for even our perception of the material is conditioned by a technics of perception.[15] As I argue in this book, a concept of the virtual has frequently been defined by how human perception has been augmented by technological enhancement. The concept of the virtual has been immeasurably productive for the study of the media, but it has also constrained inquiry. New media research has been founded on an endemic exclusion of a concept of the material with the virtual, mainly understood as the conceptual obverse of the material and the material perceptible only in its disappearance. To move beyond oversimplified accounts of the disappearance of the material, we must recognize ways in which our perception of it both conditions and is influenced by our interaction with media technologies. A substantive understanding of the material can reveal previously overlooked ways to investigate the media.

Because a concept of the material has been developed in a wide variety of disciplines, including philosophy, political economy, geography and theology, it is important to put in place clear boundaries around what is meant with the term. Each field offers a version of the concept that is specific to its epistemological position, and though I draw on some of these fields, my definition foregrounds the media's role in conditioning and reproducing our interactions with the material. The next chapter develops this claim in more detail, but for the sake of clarity, it is helpful to introduce it briefly here. The material is understood here as those objects, structures, and others that comprise one's immediate surroundings. Immediate is meant in the sense of proximity; the objects, structures and others that one interacts with in physical form on a regular basis.

In the chapters that follow, I consider a range of topics that become visible with the introduction of a concept of the material in the study of modern media. The opening chapters identify and analyze a new class of interface technologies, which I refer to as the See-Through Graphical Interface. Existing research has either overlooked entirely or underacknowledged STGI's centrality to the development of modern media. I argue that this oversight is a consequence of the emphasis on a concept of the virtual in media research. The latter part of this book examines cultural practices that make use of this type of interface in order to encourage a critical awareness or understanding of one's relationship to the material. I investigate site-specific media art, specifically artists who utilize a STGI type interface (such as Augmented Reality) in order to reveal how their work has been inspired by an exploration of site and its materiality.

In its analysis of a concept of the material and the role that media and computing devices have in producing knowledge of it, I argue that this knowledge takes form in three distinct ways. First, in a recognition of the sociocultural conditions of site, or an understanding of the social or cultural conditions of a particular location; second, in an awareness of the disciplinary function of site, or site's disciplinary role in the formation of individual subjecthood; and finally, as an encounter with the materiality of site through a recognition of the formal difference between the virtual and the material. In light of the third opening, I examine cultural practices that strive for an awareness of the material in order to produce a critical recognition of the virtual as a cultural condition. To the extent that they are able to produce a critical position relative to the virtual, such practices are said to rupture the virtual.

ORGANIZATION OF THE BOOK

Chapter 2 is a theoretical introduction to the concept of the material. It begins with an analysis of the concept's endemic exclusion or erasure from new media theory. The ascendency of a concept of the virtual in theoretical and historical research on the development of new media technologies has occluded the consideration of a concept of the material. Recalling the earlier discussion of the immediate and the mediated, the material has been defined as the conceptual obverse of the virtual with the material's presence only perceptible in its disappearance. Three influential accounts of the virtual in media studies are reviewed (the history of technologies of the virtual, embodiment and informatics, and post-structuralist theories of digital media) in order to demonstrate how each is grounded in an exclusion of the material. Chapter 2 also offers a definition of the material that responds to but is not informed by these exclusions.

On the basis of this definition for the material, Chapter 3 examines how a concept of the material can provide insights into the history and theorization of media and computing interfaces and the visual logics by which they operate. I argue that it is necessary to define a new class of computer interface technologies in order to distinguish between technologies of the virtual from technologies that can enhance knowledge over the material. This class includes devices as diverse as Augmented Reality, Head-up Displays in aircraft and cars, predictive gunsights, and others. Chapter 3 also explores the history of technologies that mediate a relationship to the material by focusing on a study of the first STGI, the Mark II Gyro Gunsight, a predictive gunsight developed by the British military in the latter half of World War II and a predecessor to modern Head-up Displays. In tracing the origin of the STGI to military technologies, the chapter considers the military's ongoing role in the research and development of computing interfaces. Paul Virilio exercised lasting influence over the theorization of the visual logic of cinema and military technologies; what he has referred to as the "fateful confusion of eye and weapon."[16]

Chapters 4 through 6 extend the discussion of the STGI and its implications for media research by considering cultural practices that employ STGI type interfaces. Chapter 4 analyzes Krzysztof Wodiczko's site-specific media projection artwork and contends that Wodiczko has utilized STGIs for aesthetic investigation of our interactions with the material. While the Head-up Display and other military STGIs are aids for enhancing knowledge and control over one's material environment, site-specific media artists, by contrast, have employed media technologies as part of an aesthetic strategy to open a critical intervention into one's immediate physical surroundings. Wodiczko's 35mm slide and video projections have been characterized as "interrogative," that is, critical of social and political problems associated with site. Building upon discussions of site specificity in art, this chapter argues that his projections can be read as an interrogative address of three different conditions of site, its sociocultural conditions, its disciplinary function, and its materiality.

Jason Farman, Nick Kaye, Gabriella Giannachi,[17] and others have recently insisted that there is a gradient between the real and the simulated, the prerecorded and the live, the virtual and material. The material and the virtual do not exist in simple binary opposition. Accordingly, the STGI can be used in a wide range of hybrid applications that vary from the critical experience of the material, as in Wodiczko's projections, to an expansion of the virtual into the experience of one's physical environment. Nowhere is this more in evidence than in future-oriented game designs for Augmented Reality devices, which insert computer generated objects and characters into the view of one's surroundings.[18]

In Chapter 5 I continue to examine media art, but consider examples that reflect this range of hybridity; artists who create sited works but do not necessarily strive for a critical interrogation of the material in all the ways explored in previous chapters. I employ the term site responsive to characterize the works discussed in this chapter. Site responsive is intended as a more inclusive notion of how media art responds to location. It is a more open-ended category for media art that includes both site-specific art and work that is agnostic about site specificity. I consider three groups of media artists who offer diverse models of site responsiveness. In the work of Janet Cardiff and George Bures Miller, and the art collectives, Minneapolis Art on Wheels and Mobile Experiential Cinema, we see a growing interest in using See-Through Graphical type interfaces for the creation of immersive media experiences away from the space of the gallery or desktop computer.

Since 2003, I have been utilizing STGI technology in my own art practice. Chapter 6 introduces this work by reviewing two site-specific installations. These projects are practice-based research vehicles for exploring questions surrounding the material and the critical possibilities contained in its theorization. My practice has been inspired by an effort to draw out what I have found successful in existing site-specific artworks, specifically how they foreground the material as a response to the rise of the virtual as a cultural condition. I develop a notion of a "rupture of the virtual" that identifies cultural practices that open an encounter with the material as a critical intervention into the virtual. The work considered in this chapter strives to develop a repertoire of techniques that produce this type of intervention.

Finally, the Conclusion offers linked speculative claims about the possibilities for the rupture of the virtual. One of the recurrent themes of the book is the potential for site specificity to open an encounter with the material and critical reflection on the virtual as a cultural condition. The Conclusion suggests that this type of critical intervention is not limited to the realm of art, rather, it can be discovered in cultural practices beyond the aesthetic. (Examples as diverse as urban farming are considered.) The rise of practices that strive for an encounter with the material can be understood as harnessing a resistance to rise of the immateriality and heightened mediation of experience. This resistance does not rely on a nostalgic return to some pre-virtual idyll, but as a critical awareness of the social and economic forces that have contributed to this condition. The Conclusion also reflects on forces poised to mitigate the power of site specificity in art, in particular, the rising popularity of outdoor media and cinematic projection in urban spaces. I urge caution in any confusion of outdoor projection with the kinds of site-specific projection considered in this book. When not deployed in a way that realizes the critical potential of site specificity, outdoor media threaten to project the spectacle out into the materiality of public space.

A New Materialism for New Media Studies

INTRODUCTION

When thinking about the media, there is a tendency to focus on those elements that distinguish them from other forms of representation: that they cast light and sound, the immateriality of their screen-based images, and the constructed environments in which we make ourselves captive to these images. Stanley Cavell's poetic reflection on cinema distilled an early fascination with the philosophical and phenomenological questions posed by our interactions with the media:

A screen is a barrier. What does the silver screen screen? It screens me from the world it holds—that is, makes me invisible. And it screens that world from me—that is, screens its existence from me. [19]

Cavell's lines were penned in reference to the magic of cinema and its immaterial imagery that draws us imaginatively into the world on the display, but keeps us at arm's length from it. His sentiment that the screen is a barrier prefigures an early definition of the virtual, as the "liminally immaterial."[20] The virtual world depicted is a space of illusion that does not have physical form outside of its immaterial existence on the screen.

In contrast to Cavell's Cartesian musings about the screen, media artists have recently been experimenting with emerging visual technologies in an effort to explore their capacity to unsettle our perception of our immediate surroundings. Mateusz Herczka and Pär Frid's new media art project, *Reverse Avatar* (2010), for example, challenges the manner in which we relate to our physical environment through media interfaces. The basis of the work is a piece of technology that Herczka and Frid designed: a set of off-the-shelf video goggles that is modified to receive a live video feed transmitted wirelessly by a remote camera operator. When the camera operator stands directly behind viewers wearing the goggles, they can see themselves moving through physical space (see Figure 1 and 2). The live feed it provides is reminiscent of the disorienting perspective frequently adopted by cinematographers (think, Gus Van Sant's extended steadicam shot through high school hallways in *Elephant* (2003)) and first person video games. But in *Reverse Avatar*, viewers are seeing themselves in the position of the protagonist. The interface has a derealizing effect on viewers' perception of themselves not only because they are interacting with their immediate surroundings through a live video feed, but because of the out-of-body perspective the device affords: viewers see themselves seeing the world—the protagonist in their own live cinematic drama.

As Herczka and Frid's work illustrates, screens no longer just screen us from the world. Screens can screen our world, that is, the world immediately around us. The screen is not always an opaque barrier, but can be a media-enhanced window onto our surroundings. As I establish later in this chapter, in contemporary research on emerging interface technologies, like Augmented Reality, where attention is being paid to how the boundary between the computer generated and the real are being blurred (due to the arrival of mobile computer interface technologies),[21] a concept of the virtual continues to predominate. The virtual guides theoretical and historical research into interactive, immersive new media technologies, but this research is founded on conceptual exclusions that have limited the breadth of how we understand new media.

This chapter addresses new media theory's endemic exclusion and erasure of the material. It has been precisely through its conceptual erasure that the material has served to draw attention and research to a corresponding rise of the virtual. To employ an idea drawn from deconstruction, the material has been "supplemental"[22] to the celebrated definition of the virtual, and the material has yet to be theorized apart from its contingency. This has limited the ways in which to conceive of the media's role in shaping one's knowledge and awareness of the material. An examination of the concept can open new pathways for media research, that include substantive ways to rethink the history and function of the computer interface and the existence of media and cultural practices that enable us to recognize and reflect upon the materiality of our physical surroundings.

I demonstrate this claim by examining three influential accounts of the virtual in which an idea of the material has had a role in constituting knowledge precisely through its exclusion. These three accounts of the virtual include post-structuralist theories of digital media, history of technologies of the virtual, and embodiment and informatics. This is only a selection of the many definitions for the virtual,[23] but I would not go so far as Anne Friedberg, who claims that the term "has lost its descriptive power,"[24] because of the proliferation of meanings surrounding the term.[25] Rather, the influence of the concept is evident precisely in its ubiquity today. The virtual is used across a wide spectrum of inquiry, each definition reflecting the nuances of the approach. Given this diversity, it would be fruitless to try to generalize across them all. The intention here is neither to collapse the distinctions between the various definitions for the virtual, nor to imply that this reading applies to all of them without regard to their differences. The choice of these three influential accounts of the virtual illustrates the existence of a pattern of supplementarity established early in new media research, one that still exerts influence over the use of the concept today.

On the basis of this analysis, a definition of the material is offered, one that responds to, but is not informed by, these exclusions. The material, like the virtual, is deeply embedded in many areas of speculation, from political economy to philosophy and theology. The conceptualization developed here resonates with some of those found in other fields, but these connections are not explicitly explored.

HAUNTOLOGY OF THE REAL: THE VIRTUAL AS GHOSTLY SUPPLEMENT

The virtual made an early appearance in post-structuralist and deconstructive accounts of the rise of new media technologies and attendant cultural and economic transformations. In *Specters of Marx*, Derrida links the virtual to our experience and conception of the real, specifically the manner in which the virtual constitutes our understanding of the real, but is overlooked in our perception of it.[26] This absence constitutes a gap in our understanding of the real; a gap, which in the course of his investigation, is unveiled as a blind spot that configures knowledge in Western thinking. In what follows, I reveal how, despite the recognition of the role of the logic of supplementarity in the theorization of the virtual, the definition of the virtual itself was contingent on an exclusion of a concept of the material.

To gain insight into the excluded status of the virtual, Derrida associates the virtual with ghosts, those immaterial beings that haunt the living yet are excluded from them. "What is a ghost? What is the effectivity or the presence of a specter, that is, of what seems to remain as ineffective, virtual, insubstantial as a simulacrum?"[27] Despite their marginal status, Derrida observes that Shakespeare's ghosts are frequently characterized as being in possession of knowledge or understanding that surpasses the everyday: "A specter... is unreal, a hallucination or simulacrum that is virtually more actual than what is so blithely called a living presence."[28] As ghosts have the ability to upset the natural order of things because of their possession of knowledge that surpasses the living, the virtual, when acknowledged as integral to the constitution of knowledge about the real, similarly has the potential to destabilize what we understand about reality.

Derrida destabilizes the assumption that the real is the locus of activity and meaning with his insistence that the immaterial forces of new media are increasingly central to the formation of the real. He associates the virtual with a variety of contemporary developments, including "spectral effects, the new speed of apparition (we understand this word in its ghostly sense) of the simulacrum, the synthetic or prosthetic image, and the virtual event, cyberspace and surveillance, the control, appropriations, and speculations that today deploy unheard-of powers."[29] The nature of electronic information is an immateriality that is becoming central to contemporary social, economic and cultural developments in the West. For Derrida, the coinage "virtual reality" best captures the nuances found in his redeeming of the concept of the virtual. It is not just a technological device that enables the creation of immersive computer worlds. The neologism also captures the virtual's relationship to the real in contemporary society: reality is possessed by virtual processes to such an extent that the virtual has come to govern developments in the West. The virtual is a supplement to the real in that the material is integral to it, but has been excluded from consideration.

Derrida's characterization of the virtual is emblematic of a variety of postmodernist readings of contemporary media culture during the period. In his review of this work, Mark Poster compares Derrida's and Jean Baudrillard's respective positions on the virtual in order to highlight their preoccupation with the concept.[30] In contrast to Derrida, Baudrillard's reading of virtual reality as a metaphor for our cultural condition is an extension of his earlier ideas on simulation. Without going into too much detail, Baudrillard contends that electronic media, consumer capitalism and commodity culture have come to replace religion and other traditional sources of value that once grounded truth and meaning. Because it retains an antiquated sensibility in which words had stable or fixed referents, Baudrillard rejects the term "reality" as an accurate characterization of postmodern culture. With the triumph of electronic media and consumer capitalism, simulation is preferable to reality, because signifiers have become unhinged from their referents. In this context, Baudrillard sees virtual reality as nothing less than the technological realization of simulation.

With Virtual Reality and all its consequences, we have passed over into the extreme of technology, into technology as an extreme phenomenon. Beyond the end, there is no longer any reversibility; there are no longer any traces of the earlier world, nor is there even any nostalgia for it.[31]

Because virtual reality environments are constructed by the hands of human designers, all experience within them is the playful construction of postmodern culture. In virtual reality, simulation becomes reality; the word made immaterial flesh.

In both Derrida and Baudrillard's work, the concept of the virtual is overdetermined, referring to a wide swath of epochal economic and cultural changes. Their versions of the concept destabilize prevailing conceptions of the real that do not take into account the epistemic transformations occurring with the virtual's rise. They seek to dispense with a materialist understanding of the world in which meaning is associated with the appearance of phenomena. For Baudrillard, the claim that all reality is virtual reality is based in his insistence that reality is constructed out of virtual, immaterial processes, including the flow of electronic images, commodity culture and financial capitalism. Because it overlooks immaterial processes, a materialist understanding of the world does not recognize how reality has been dematerialized with the advent of simulation. For Derrida as well, the virtual is the supplement to the real in that our notion of reality is founded on the endemic exclusion of the immaterial. Derrida undermines a materialist position with his insistence that the real is always already determined by the virtual. But because virtual processes are immaterial and thus hidden from view, they remain excluded from intellectual speculation.

Derrida and Baudrillard shift attention to the virtual as the locus of activity that constitutes meaning. Next, I consider two additional concepts of the virtual that were influenced by post-structuralism's work. Though they differ in terms of their approach, the accounts share a preoccupation with foregrounding a concept of the virtual. They demonstrate how the virtual has been integral to social, cultural and economic transformations in which we are in the midst. As a result, the virtual comes to define new media research and is no longer peripheral to their study.

THE HISTORY OF TECHNOLOGIES OF THE VIRTUAL

In the years since Derrida and Baudrillard's speculations on the virtual, theorists have refined the concept's usage to such an extent that it no longer aspires to characterize epochal cultural or philosophical shifts. An influential version of the concept is found in the description of perceptual changes and technological innovations that have been involved in the rise of interactive, immersive media environments.[32] The virtual has been applied to the construction of computer-generated environments, such as video games and Virtual Reality, but it is also located in the rise of panoramas, camera obscuras and other early media devices integral to the development of contemporary spaces of immersion. In her historical study of the rise of the virtual, Anne Friedberg charts its history from Renaissance perspective devices to today's Graphical User Interface (GUI). Renaissance painters used enframing devices, a kind of picture frame, as a window onto physical space as an aid to painting according to the laws of perspective. For Friedberg, today's GUI finds its origins in these Renaissance devices in that the GUI has become a metaphorical window onto information space, a virtual space that no longer needs to obey the laws of perspective.

If we venture a different look at this history and consider the status of the material, we can recognize how the material has been elided from accounts of technologies of the virtual. In this history of new media, there has been an emphasis on the historical and perceptual development of the virtual in the creation of ever more sophisticated immersive and interactive mediated spaces, which

Friedberg and Grau refer to as "spaces of illusion." Their historical work addresses the separation of the virtual from one's physical environment in technologies, like the GUI or video games, in which code comes to "define the entire informational or perceptual environment for the user."[33] Though it draws on the physical world for metaphors (such as the window, the desktop, etc.), the GUI does not have to correspond to our surroundings or the laws that govern them. These studies of the virtual frequently culminate in an extended examination of the promise and potential of Virtual Reality, which is undersood as further evidence of the concept's primacy. VR is read as the realization of what was inherent in earlier media forms: the possibility of the creation of fully immersive and interactive virtual worlds and the supersession of the real by the virtual.[34]

Consider again Stanley Cavell's philosophical musings quoted earlier. I suggested that they be read as an early definition of the virtual with Cavell's emphasis on the screen as a barrier. The "world" to which he refers is one of cinematic projections—a virtual world created out of imagery projected onto the silver screen. There is, however, a crucial ambiguity in his use of the word, identifiable now that we've considered the status of the material in the literature on the virtual. It could also refer to the world outside the theater, that is, one's physical environment. Here, in the darkened room of the theater, that which is screened from me is the world that I inhabit, the material world.

This ambiguity reinforces a claim about the virtual suggested earlier: screens capture our gaze in fixed attention. So drawn into this immaterial world, we are screened from our surroundings on the other side of the projections' borders and theaters' walls.

The virtual has been instrumental in promoting an understanding of the perceptual and technological changes that have taken place with the rise of modern media, but only those technologies that fit into this particular account of the virtual have been included. In other words, the virtual has been instrumental in producing a kind of teleological thinking that sees new media's history written from the vantage point of the virtual. It is a history premised on the leading role that the virtual has had, which organizes historical and theoretical insights around it. We can see this tendency continuing in research on emerging technologies. In recent investigations of mixed reality interfaces, for example, there is a tendency to conceive of interfaces like Augmented Reality as an outgrowth of earlier technologies of the virtual. Jason Farman notes how in Augmented Reality the material is subsumed by a virtual or informational dimension.[35] In Augmented Reality and related mixed reality type interfaces, one's physical surroundings become an "information interface"[36] in that one's geographical location becomes an indexical marker to online information accessible by mobile computing devices. Augmented Reality is an interface technology that enables the material to be colonized by information through the transformation of the material into a marker of virtual information. In that the virtual is privileged over other possible explanations, this characterization is evidence of the continuing emphasis on the virtual in research on contemporary interface designs.

EMBODIMENT AND THE MATERIALITY OF INFORMATICS

In contrast to the two areas reviewed above, the research on new media and embodiment applies a different set of theoretical assumptions in its work on the virtual. There have been two interrelated positions in this area of research: the claim that our interactions with digital communications lead to a kind of disembodiment because of the immateriality of information, and research on how the body and information technologies mutually constitute each other, because information technologies extend humans' perceptual capacities. Though they differ in terms of how they understand the virtual, what they share is a conception of the material as coextensive with the human body at the exclusion of a concept for the material as outside of the perceiving subject.

In her account of the former position, Emily Apter argues that we leave our bodies behind (that is, at the computer terminal) in online interactions. For Apter and others, this disembodiment is potentially liberatory, because we can free ourselves from the rigid markers of the body's constraints (i.e. race, gender, disability). This celebration of disembodiment is rooted in the belief that humans could escape the essentializing forces that construct the body through the use of digital technologies. Digital technologies can enable a "postidentitarian politics rooted in the obsolescence of racial and ethnic categorizations."[37]

While Apter depicts the body and informational systems as in opposition, Katherine Hayles has posed a more dynamic way to conceive of this relationship. Instead of portraying technology as a disembodying force, Hayles has insisted that human and information systems are in a co-evolutionary dance, which she has referred to as an "intermediating dynamic."[38] In earlier work, especially "The Materiality of Informatics," she offered a preliminary version of this idea with her insistence that the human body exists as a kind of constraint on the formation of information systems, because technologies are only adopted when they can be integrated into existing human functioning through a dual process of "inscription" and "incorporation."[39] She uses the example of the development of audiotape with its capacity to record and objectify the human voice. Its adoption "changed the relation between voice and body" producing a "new kind of [posthuman] subjectivity" that she identifies in examples of modernist fiction and poetry.[40]

Hayles' insistence that information systems only develop within the constraint of the materiality of the body has influenced research in this area. It has shifted discourse to ways in which the human body is a site for the inscription of informational systems. In *Materializing New Media*, for example, Anna Munster explicitly relates the two in her observations about the embodying tendencies of new media.[41] Munster observes the manner in which the body, rather than excluded or disembodied in one's use of various new media technologies, is directly involved in the interaction. Indeed, Munster uses the term material interchangeably with the body and the corporeal. The "materializing" of new media (to which the title of Munster's book refers) is precisely the inclusion of the body's physio-motor capacities or physical presence in new media interactions.

In his study of embodiment and new media, Mark Hansen has traced this interest in the body to a more general "turn to the body" in cultural theory, culminating in the 1990s with the development of the "cultural constructivist paradigm."[42] The adoption of cultural constructivism in feminist and queer writings on technology then gave rise to the figure of the cyborg reclaimed as a synecdoche for emancipation from the reigning essentialisms that dominate racial, ethnic, gender and sexual categories. Despite characterizing her manifesto as an "ironic political myth,"[43] Donna Haraway's fantastical depiction of the cyborg in "A Cyborg Manifesto" accelerated into a proto-utopianism surrounding late twentieth and early twenty-first century information and communication technologies. From the Internet to cellphones to prosthetic enhancement of the body, new digital technologies became the site of growing fascination, because they were perceived as enabling a liberation from the constraints of the everyday. Work, like Apter's, is born from a belief in information and communication technologies' liberatory potential.

Hansen shares a conception of the body that renders it as a raw materiality that is given extension or enhancement with information and communication technologies. But dissimilar to cultural constructionists, Hansen grounds his work in an updating of a phenomenological account of perception, specifically a reassessment of Merleau-Ponty's notion of the "technics" of human embodiment. For Merleau-Ponty, technics refers to any technologies that enhance human perceptual capabilities. Hansen's approach to human perception is an effort to illustrate how technology is "always already embodied,"[44] that is, human perception always already bears the mark of technological enhancement. The body's perceptual capabilities are the frame within which media technologies are developed and integrated into human perception.

The reason for the interdependence of human perception and media technologies that enhance it derives from the difference between the body and the realm of sense data, two phenomena that differ in form. According to a neo-Bergsonian account of perception, sense data is the realm of the virtual, because visual and auditory stimuli are immaterial when they enter perception. Sense data becomes "actual" or "real" by means of human perception, which attends to and fixes images in memory.[45] The body, it can be said, gives reality to the virtuality of the image through embodied perception. Put differently, there is a gap between the body and the virtuality of sense data which needs to be bridged in order for sense perception to occur. Media technologies, such as film, computers, and virtual reality, bridge this gap by extending the body's perceptual capabilities into the realm of the virtual.

Thus, when Hansen writes that "virtual reality comprises something of a reality test for the body,"[46] he is referring to this difference between the materiality of the body and the virtuality of sense data. Virtual reality is a reality test for the body, because visual images are given actuality by the perceiving subject. In positing the body's difference from the virtual in this way, Hansen's work is comparable to other theories of embodiment that conceive of the body and new media as in opposition, with the body the domain of materiality and media that of virtuality.

Slavoj Zizek's critique of cultural constructionism can help identify what is limited about the depiction of the body as the locus of materiality. Employing language familiar to deconstructionists, Zizek insists that cultural constructionism is founded on an "exclusion / foreclosure that grounds this very horizon."[47] The effort to unmask and resist all essentialist positions regarding the body forecloses the possibility of recognizing the existence of a background from which subjectivity arises. For Zizek, this excluded background is capitalism, which for the purposes here is less relevant, but by applying his argument to the role that the body has in the literature on embodiment and new media, a related critique can be made. In this work the differences between material/virtual, inside/outside, and body/environment are reified to such an extent that the subject is considered material and everything outside the subject (because it enters the body as sense data) is virtual.

Zizek's accusation that cultural constructionism fails to acknowledge the existence of a background that makes possible subjectivity is applicable to the research on embodiment in that it relies on an idea of a gap between the body and the virtual. However, this view overlooks the extent to which there is a determining background or environment that makes possible the materiality of the body. That the boundary between the body and environment is porous is familiar to cultural theory. Gender, for example, is a performative act made up of behaviors learned via sense perception. As Judith Butler has argued, the border between the inside and outside is blurred, as external norms or forces constitute the individual and have determining effects on the subject's body in terms of how it expresses itself and functions as a gendered being.[48] If external norms and forces have a role in constituting the materiality of the body, then any simple opposition between body and environment disregards the complexity of the body's materiality.

Returning to Zizek's critique, we can conclude that theorists of embodiment do not properly acknowledge the existence of background forces that ground the materiality of the body. In contrast to other areas of research in new media, the research on embodiment offers a conception of the material, but one that is too restrictive. The material is portrayed as coextensive with the perceiving subject, and what is outside the subject is consigned to the realm of the virtual. This position does not recognize that the construction of the body presupposes the existence of forces external to it that ground the possibility for its materiality. In the next section, I offer an alternative formulation of the material, one that posits its presence outside the perceiving subject and foregrounds the media's role in enhancing knowledge of the material.

RE-MATERIALIZING NEW MEDIA STUDIES

Derrida's observation about the exclusion of the virtual from intellectual speculation must be reconciled with the concept's centrality in new media research. The virtual is not only presumed to inhabit the everyday, but it has a role in radically reshaping it. The concept has moved from periphery to center, from exclusion to inclusion, and continues to influence the way in which we understand media culture. In enabling researchers to trace tendencies in the historical development of new media, it has simultaneously constrained inquiry because it has been founded on an endemic exclusion of a concept of the material external to the perceiving subject. The material has been conceived of as the supplement to the virtual with the material perceptible in its disappearance. In contrast to the virtual, the material has not been given sufficiently considered as a constitutive element in our interactions with media technologies. To move beyond accounts of the disappearance of the material, its total mediation, or association with the body, we must recognize ways in which our perception of it both conditions and is influenced by our interaction with media technologies. Developing a substantive understanding of the material can open new ways to investigate the media.

In order to provide a manageable definition of the material, it is important to put in place clear boundaries. A concept of the material has been utilized in a wide variety of disciplines, including philosophy, political economy, geography and theology. Each field, of course, offers a version of the concept that is particular to its epistemological position. The definition offered here does not try to integrate these diverse approaches, nor even explicitly engage with them. Though it draws on some of these fields, this definition is appropriate to the study of the media, one that foregrounds the manner in which the material both conditions and is produced in our interactions with the media.

The material is understood here as the presence of objects, structures, and others that comprise one's immediate surroundings. Immediate is meant here in the sense of physical proximity, or the objects, structures and others that are within the range of one's perceptual senses. This can include things that are perceived with the aid of devices that augment sensory perception, such as eyeglasses, telescopes, megaphones and shotgun microphones, provided they can also be experienced without augmentation with minimal effort. An example of an experience that is immediate, but augmented, is the use of high-powered binoculars to look at trees at the edge of one's physical surroundings. Because one could move closer to them in order to see them without the aid of the binoculars, it is included in the definition of one's immediate environment. By contrast, video chatting with friends on the other side of the world might give one a powerful feeling of closeness to them, but they are not physically proximate without great effort and time.

The latter example brings to mind an important distinction: physical proximity is often conflated with, but should not be confused with, perceived or felt closeness. It is important to distinguish between them conceptually. Felt closeness does not require physical proximity. One can feel an attachment to a place, for example, despite the fact that one may never have visited there. The difference between proximity and closeness employed here is a technical distinction not an evaluative one –what is not being suggested is a preferability of one over the other. The media have a power to shrink one's sense of distance and time to such an extent that one can feel close to things and others who are far away, but the definition here strictly concerns physical proximity.

A related problem exists in recent intellectual efforts to reclaim the material. In fields as diverse as anthropology, geography, and others, there has been a call to return to a material analysis of things and objects, in order to distance research from the perceived excesses of the social and cultural constructedness of phenomena. Sometimes referred to as the "materialist return,"[49] this position skirts close to a longing for an immediate, lived experience of the world that is oftentimes nostalgic for a perceived closeness of objects and environment. A problem with the materialist return is its epistemological basis. This sense of closeness is grounded in a presumption that meaning emanates from the objects and things themselves. Critical of a similar tendency, the nineteenth-century German philosopher Arthur Schopenhauer took to task a comparable materialist position, one in which "everything objective, extended, active, and hence everything material" is regarded "as so solid a basis for its explanations."[50] Schopenhauer is suspicious of this position for it imagines the material as a ground from which truth derives with materialist philosophy's task to uncover this meaning.

The definition of the material offered here, however, does not assume that the material is the ground of meaning. Like the virtual, the material is embedded in diverse systems of signification and is the confluence of complex forms of meaning, histories and cultural legacies of which one may not be aware. These systems of signification constitute the way in which one evaluates, understands, sentimentalizes and relates to one's physical surroundings. The media is one system of signification that has a role in constraining and enabling one's experience and understanding of the material. The material is defined here as that which is in our immediate physical surroundings, but the forces that condition the material itself can be non-local and immaterial.

In order to develop the significance of this definition for the study of new media, the next chapter turns to an examination of the role that computing interface technologies have had in enhancing knowledge over the material. I focus specifically on a class of interface technologies I refer to as See-Through Graphical Interfaces, computing interfaces that overlay graphical information into one's field of vision, and the reasons for distinguishing them from technologies of the virtual. Instead of opening up windows onto virtual spaces, See-Through Graphical Interfaces enhance knowledge over things in one's immediate physical surroundings.

1
Mateusz Herczka
and Pär Frid,
Reverse Avatar,
2010

2
Mateusz Herczka
and Pär Frid,
Reverse Avatar
[alternate view],
2010

3
The Origin of the See-Through Graphical Interface

In this chapter I define a new category of media and computing interface devices, the See-Through Graphical Interface (STGI), in order to acknowledge its unique history and mode of visuality that is distinct from technologies of the virtual. In contrast to media that are a virtual window onto constructed worlds or spaces, STGIs project graphical information into the view of one's immediate physical surroundings. The last chapter detailed the legacy of a concept of the virtual in research on media and interface technologies, which has resulted in an impoverished conception of the material. By extension, this has led to the under-theorization and recognition of the historical importance of computing interfaces that enhance knowledge over the material. There are a number of computing devices, such as gunsights developed during World War II, that precede the invention of many of the hardware and software innovations associated with the development of technologies of the virtual.

Defining a new category of interfaces is required in order to reveal this history and underscore how they pose a mode of visuality that is distinct from other screen-based media. I refer to these devices as STGIs, because users are able to see the environment through the computing device; users' surroundings are the "background" on which computer-generated or virtual objects are projected. This category includes technologies as diverse as Head-up Displays, Augmented Reality, military gunsights and others. With the arrival of interfaces like Google Glass, especially, there has been a growing body of research about these technologies.

BETWEEN THE VIRTUAL AND THE MATERIAL

In the past few decades technologists and research scientists who work at the intersection of mobile computing, vision, interface design and computer graphics have been conceptualizing and forecasting the arrival of a wide variety of mobile interface devices, including but not limited to Augmented Reality. Milgram and Kishino, for example, exercised an early influence on the theorization of Augmented Reality with their taxonomy of visual display technologies, in which they conceive of Augmented Reality as a "subset" of Virtual Reality technology.[51] By contrast, Ron Azuma, a pioneer and innovator of Augmented Reality, locates AR on a spectrum ranging from the completely synthetic (Virtual Reality) to the completely real (telepresence) with AR situated between these two poles.[52] Steve Mann has been developing and using computing devices that technologically enhance his vision since the 1970s. As distinct from the others, Mann prefers the terms mediated reality, which includes both virtual and augmented reality. For Mann, mediated reality is a term that emphasizes the use or attributes of the media interfaces instead of their technical details.[53]

There is little consensus among technologists either about the future direction of these technologies, or the concepts and terms to describe, contextualize and analyze Augmented Reality-type devices. For their part, media and visual studies researchers have offered limited help in this regard and have generally been slow to acknowledge their distinctiveness from other digital technologies. One reason for this is what Nathan Jurgenson has referred to as a prevailing "digital dualism" within scholarship that has constrained thinking about digital media. This is a tendency to render digital processes and the physical world in a binary opposition with digital technologies conceived as enabling virtual experiences (i.e., immaterial, disembodied, digitally fabricated), while the physical world exists separate from the virtual. According to Jurgenson, this has resulted in the under-theorization of computing technologies that mediate our relationship to our physical surroundings, like AR, for they pose a mode of visuality distinct from existing technologies of the virtual.[54]

While the virtual, broadly conceived, concerns media that involve a distancing, if not a complete separation, from one's material surroundings, with AR the material figures importantly in one's interaction with the medium. If we look more closely at how some researchers speculate AR will function, we can recognize how the relationship between the material and the virtual is more complicated than it at first appears. AR engineers foresee the development of computer vision and recognition algorithms (a development that is years from realization, if the technical challenges are ever surmounted) that would enable software to recognize objects, people and spaces in one's material environment.[55] Such a system would require that a user's material environment is encoded, organized, and made subject to algorithmic analysis.

Lev Manovich identified "numerical representation" as one of the main principles of new media, a concept that can be usefully applied to this analysis.[56] According to Manovich, what is distinctive about new media is their ability to represent or encode analog information or processes as numerical (i.e., binary) data. While Manovich had in mind the migration of analog media to digital formats (e.g., digital photography, digital audio, and digital video), his insight also illustrates an important dimension of AR. As maintained above, AR engineers are developing techniques to encode one's material surroundings so that they can be subject to computer-based algorithmic analysis and interpretation. In Manovich's terms, in order for such a system to be feasible, the numerical representation of one's material surroundings is required. Users access this data, when retrieving information about their environment through an AR device. In this way we can understand AR as the expansion of the virtual into the material, insofar as users' material surroundings are encoded as a numerical representation, and interactions with the environment are mediated by this representation.

Jason Farman's *Mobile Interface Theory* is a forward-thinking account of the informational representation of physical surroundings in interface technologies, like AR, and it is an example of work that strives to move beyond the pervasive dualisms of which Jurgenson is critical. Farman considers a number of different kinds of mobile media technologies, including Augmented Reality enabled smartphones, SixthSense, and locative games, and their effect on our experience of the physical world. In his analysis of Augmented Reality specifically, Farman notes the growing interpenetration of material space by an informational or virtual dimension. He writes:

The move from personal computing to pervasive computing, a shift characterized by the move from immobility to mobility, has allowed for online space to interact with material space in unprecedented ways.[57]

Farman notes an erosion of the distinction between the material and the informational with the rise of mobile computing. Physical surroundings are an "information interface," because one's geographical location becomes an indexical marker to online information accessible by mobile computing devices.[58] While technologies of the virtual entail the creation of environments that have no necessary corollary to the material world, with AR a distinction between the material and its numerical representation is integral to users' interaction with the interface. The presentation of computer generated graphical information in AR constitutes an information layer overlaid onto the experience of the material environment. Users of AR devices still move through and interact with their material environment, which the interface facilitates.

Though it is prescient in its analysis of the perceptual changes taking place with new interface technologies, Farman's argument is contingent on the rise of mobile digital computing and the effect that it is having on our experience of the material environment. This is, however, too limited an account of the history of these technologies. In order to make sense of them, this article situates AR as part of developments dating back as far as World War II with the invention of the first STGI, a predictive gunsight found in military airplanes. A precursor to modern Head-up Displays and AR, the gunsight provided users with an information layer in the experience of their physical environment, which enhanced knowledge over their surroundings. The next section begins with an overview of the military's ongoing role in the research and development of STGIs for the technological enhancement of vision, a research program that introduces the case study of this chapter.

AUGMENTED REALITY AND THE LAND WARRIOR PROGRAM

In July 2009, soldiers of the 5th Brigade, 2nd Infantry Division of the U.S. Army were deployed to Afghanistan under President Obama's call for a troop surge to counteract a rapidly deteriorating military situation in the country.[59] What was noteworthy about this division was that each soldier was outfitted with the latest in military hardware and armament, a suite of technologies known as the Land Warrior system. The Land Warrior system has been in research and development since 1993, when the military made an initial call for proposals for contractors to develop different components of the weapons system.

The call described the Land Warrior system in the following terms:

> The LWS [Land Warrior System] will be used by the dismounted war fighter at the tactical level of war. The LWS will include an integrated complement of weapon, protective clothing and headgear… to enhance his lethality, command and control, survivability, mobility, and sustainment.[60]

As Chris Gray observed in his chapter on the "Cyborg Soldier" in *Postmodern War*, the military has long experimented with new warfare technologies, with the soldier's body frequently the site of prosthetic enhancement and experimentation.[61] From devices that augment a soldier's sight and communication abilities to drugs that prevent stress and increase endurance, military research programs have aimed to augment a soldier's lethality and reliability in combat through integration with technology. According to Gray, in this sense, the soldier has long been a human-machine hybrid, or "a cybernetic organism (cyborg) model of the soldier that combines machine-like endurance with a redefined human intellect subordinated to the overall weapon system."[62] The Land Warrior program fits squarely within the military's "cybernetic" research programs of the type about which Gray writes. One part fantasy: the cyborg soldier is pure ideology, an expression of the military's faith in technology's ability to enhance the mobility, lethality and commandability of combat troops; and one part reality: an active program that has the material result of channeling billions of taxpayers' dollars into university, industrial and military research.[63]

With the headgear subsystem of the Land Warrior program the military continues its longstanding influence on the research and design of computer devices for the technological enhancement of military troops.[64] According to General Dynamics, the chief contractor on the subsystem of the Land Warrior project, the headgear consists of a Head Mounted Display (HMD) attached to a mobile computer, running specialized software and connected to an array of sensors (see Figure 1).[65] The display has multiple functions: with its integrated GPS, soldiers can call up their precise geographical position on a digital map and get information on the positions of enemy targets and friendly troops. In a realization of Paul Virilio's observation about the linked developments of military and cinematic modes of visuality—what he has referred to as "the fateful confusion of eye and weapon"[66]—the HMD is a remote display for a camera mounted on a gun, allowing soldiers to aim their weapons without looking through the gunsight. Soldiers literally see from the perspective of the firearm, as if they were their own gun. The HMD can also connect to a night vision camera, allowing soldiers to see in the dark, and with the aid of thermal imaging sensors, even to look through walls. Lastly, the HMD is an integrated communication device; it enables superiors to transmit data to soldiers in the form of digital images, video, data messaging, and other kinds of combat information, keeping soldiers up-to-date on battlefield conditions.

In this way, the HMD is a repository of information from a variety of sources, including commanders, other troops, and surveillance data from satellites and airborne sources, connecting soldiers to a real-time battlefield information network. The soldier, in effect, becomes a node in an information network and can be remotely controlled according to changes in the combat situation.

While the Land Warrior HMD now functions like a highly mobile computer display, the U.S. government has been funding an active research program in Augmented Reality in an effort to craft the next generation of the interface.[67] The HMD currently designed for the Land Warrior system basically functions like a sophisticated version of the eyepiece of a video camera; the display takes up the entirety of the users' field of vision, allowing no unmediated visual access to their surrounding environment. An AR-enabled HMD (or AR-HMD), on the other hand, facilitates interaction with users' immediate physical surroundings. Users are able to see their environment through the headset. Information is superimposed directly onto the battlefield itself, allowing soldiers to access a wealth of battlefield information, such as the locations of friendly and hostile forces, improving their understanding of their geographical position and the potential dangers in the environment. By projecting computer-generated or virtual objects into soldiers' field of vision and thus onto their view of their physical surrounds, the AR-HMD is said to improve "situational awareness" by providing battlefield information to soldiers in a less distracting manner.[68] Soldiers are able to keep their eyes on their material surroundings when retrieving information, lessening the dangers of being distracted while in combat.

When the Land Warrior system is situated in the context of the military's longstanding role in developing computing technologies,[69] the military's foray into display interfaces should come as little surprise, for the AR-HMD is deeply connected to earlier warfare technologies used for graphical interface and visualization purposes. There are clear connections between HMDs for foot soldiers and the Head-up Display (HUD) found in the cockpits of many military and commercial aircraft.[70] The HUD is an interface device that projects graphical information into users' view of their material surroundings. For pilots, the HUD makes available real-time combat and navigational information on a transparent screen. With the Land Warrior system the miniaturization of computing and display technology has enabled the military to re-imagine the role of the HUD on the battlefield; computing devices that were once restricted to the cockpit of an airplane can now be carried on the body. In this sense, the AR-HMD is a wearable version of the HUD. Both the HMD and the HUD are computer interfaces that enhance a users' knowledge of their material environment.

THE MARK II GYRO GUNSIGHT: A PREDECESSOR TO THE HEAD-UP DISPLAY

It is difficult to determine when precisely the term was coined, but in the decades following World War II, military researchers and personnel employed the term Head-up Display to refer to a variety of interface devices to improve the visual presentation of flight and targeting information. In 1958, J.M. Naish,[71] a British research scientist for the Royal Aircraft Establishment at Farnborough (referred to as R.A.E. Farnborough), the premier military aircraft research facility through most of the early twentieth century, penned the earliest written definition of the term:

The essential feature of the pilot's instrument presentation is the ability to provide certain information from the so-called "head-down" display of flight instruments, without the pilot needing to deflect his attention from the head up position... With the difficulties of providing safe automatic control at low level this "head-up" display must as a minimum give a clear unambiguous flight director display that can be used for long periods without fatigue.[72]

Naish was well-positioned at R.A.E. Farnborough to work on the HUD, for the facility was at the forefront of the research and development of predecessors to the technology. In a series of technical papers that he wrote while at R.A.E. Farnborough, Naish proposed inventions crucial to the development of a flight-ready Head-up Display. These included a "superposed projection from a bright cathode ray tube"[73] or a display which the pilot sees "'out of the corner of his eye' using the para-foveal region of the retina."[74] According to Naish,[75] these designs were basically enhancements to existing gunsighting technologies used in the cockpits of military aircraft.[76] Naish's acknowledgement of this connection links his work on the HUD to earlier World War II military technologies such as the aforementioned Mark II Gyro Gunsight.

The Mark II was developed in 1942, and production orders were taken in that year.[77] By 1944 the gunsight was widely found in the cockpits of Spitfire and Hurricane aircraft.[78] Elements of the gunsight were the outcome of technological developments going back decades,[79] and many of the innovations used in the device were found in other combat technologies during the period. The Mark II, however, was the first STGI, because it was the first "computing" device to employ a projection screen on which graphical information was displayed that enhanced operators' performance at their task.[80]

For fighter pilots, the task is shooting down enemy aircraft, and for this purpose the Mark II performed startlingly well. The gunsight was painstakingly tested, studied and analyzed and found to dramatically improve targeting accuracy. One popular account of the gunsight summed up its efficacy in the following way: "Combat results, reports the British Air Commission, show that the fighter aircraft are now destroying nearly twice the number of Luftwaffe aircraft since the introduction of the new gunsight."[81] It is impossible to assess the factuality of these claims, but it was an opinion shared by military staff as well. Writes Trafford Leigh-Mallory, the Air Marshal of the Royal Air Force, about the Mark II:

During nearly 1200 actual attacks against enemy aircraft [the study of the Mark II Gyro Gunsight] has shown that 21% resulted in complete destruction of the enemy aircraft... The general introduction of the sight into squadrons should result in a marked improvement in the standard aiming accuracy by fighter pilots in combat with both enemy fighters and bombers. I consider that

Mark II Pilot's Gyro Gunsight to be potentially one of the most valuable air fighting assets which has been developed during the war, and I am sure that its introduction throughout my command is a pressing need.[82]

The "pressing need" for the gunsight was a dramatic increase in the complexity and speed of warfare during World War II, especially in aerial combat. Prior to World War II, pilots were trained to target enemy aircraft by making mental calculations on when and where to fire. During World War I, for example, pilots would bomb "low and slow," that is, at relatively low altitudes and slow speeds,[83] because the runs could depend on a bombardier's intuition and training. Intuitive bombing and gunnery was effective, partly because aircraft were slow moving. World War II saw the introduction of the British Spitfire and the German Messerschmitt Bf109s, aircraft that could fly in excess of 300 M.P.H.; due to their improved speed and maneuverability, pilots could no longer reliably make mental calculations on when and where to fire at enemy aircraft.

What was needed, then, was a device that could take the guesswork out of aiming, one that could automatically compute the information needed for accurate targeting of moving aircraft. The Gyro Gunsight did just this. In aerial gunnery, there are primarily three ballistics challenges in targeting: lead or "angle of deflection" firing, bullet drop, and range. The angle of deflection is the amount of "lead" needed to fire bullets that could intercept a moving target. When firing at a moving target, one needs to shoot in front of the target, or account for the lead time, in order to hit the target. In aerial dogfights, however, not only is the enemy moving, but the firing aircraft as well, which adds a layer of complexity and increases the kinds of error in estimating when and where to fire. The Mark II Gyro Gunsight calculated the amount of angle of deflection, bullet drop and range needed to hit a moving target, and it did this through the presentation of a moving aiming mark, or graticule, which a pilot lined up with enemy aircraft (see Figure 2).

It is worth pausing here to consider in more detail how the gunsight aided targeting through a study of its construction and operation. The Mark II consisted of three basic elements: the "computer unit,"[84] the graticule, and the reflector screen (see Figure 3). The "computer unit" —the brains of the gunsight— was an analog computing device consisting of a gyroscope and a series of electromagnets, which provided the offset information for the graticule. The principle behind the use of gyroscopes for predictive targeting was well established by this period. A gyroscope resists an airplane's axial movement, and by connecting the moving graticule to the gyroscope, the graticule would respond to an airplane's movement.[85] Pilots could dial in the type of aircraft against which they were fighting, which would adjust for the enemy's speed by changing the strength of the four electromagnets aligned around the gyroscope. These electromagnets account for range and gravity by offsetting the movement of the gyro.

The reflector screen consisted of a glass panel about 4 inches tall by 2 inches wide. Pilots looked through this screen when aiming their guns. A light bulb housed within the computing unit back lit the moving graticule and projected its image onto the transparent screen. When looking through the reflector screen, pilots would see an image of the graticule superimposed into their line of sight. By aligning the graticule with the enemy aircraft, pilots would have the proper amount of lead time computed for deflection firing. The predecessor to the Mark II (the Mark I Gyro Gunsight) did not utilize a reflector screen. Instead, the sighting device was a peephole, comparable to scopes commonly found on rifles today. When Virilio noted the similarities between the technologies of cinema and warfare, it was on the basis of comparisons between the peephole sight found in bombsight technologies (such as the Norden bombsight) and the eye piece of a movie camera. In this, the Mark I Gyro Gunsight was similar to other gun and bombsights during the period in its use of a peephole sight.

The transition from the peephole sight to the Mark II's reflector screen directly relates to the thesis of this chapter regarding the insertion of a layer of graphical information into the pilots' experience of their material surroundings. The reasons for the redesign concerned the restrictive aperture size of the peephole, which forced pilots to reposition themselves to operate the gunsight. Pilots had to lean forward to look through the sight, which took them out of their normal flying position, making them less prepared for evasive maneuvering and, thus, more susceptible to counterattacks. This problem is embodied in Clarke's observation about the inherent dangers in using the sight: in high speed aircraft susceptible to sudden and unpredictable movements especially during combat, there was a high potential for facial injury, as pilots had to put their face up to the peephole sight in order to target enemy aircraft.[86]

The Mark II gunsight, by contrast, addressed these issues by eliminating the peephole entirely and replacing it with a transparent reflector screen. In comparison to the peephole, the reflector sight could almost be used as a continuous extension of pilots' normal flying position. Though the sight was in no way perfect, pilots no longer had to lean forward when looking through the gunsight. They could remain in their normal flying position and operate the gunsight from there. Just as importantly, the reflector screen also improved pilots' field of view. Unlike the restricted aperture size of the peephole, the reflector screen was considerably larger and better integrated into pilots' forward field of view. The design was such that it made for a more seamless presentation of targeting information; information was projected directly into the pilots' line of sight. Put into terms defined earlier, the Mark II gunsight overlaid information into the pilots' view of their surroundings. By improving their ability to shoot down enemy aircraft, this display of visual graphical information enhanced pilots' control over their material surroundings in quite a literal way.

ENLARGING THE TRANSPARENT SCREEN

As early as February 1941, British military researchers explored the possibility of enlarging or even completely removing the reflector screens used in aircraft gunsights in order to increase the area in which a pilot could target an enemy aircraft and reduce the gunsight's obstruction of the forward field of view. In reference to the bulk and size of an early reflector gunsight, the Chief Superintendent of the R.A.E. Farnborough remarked, "the layout of cockpits in some of the new S.S.F. [Single-seater Spitfire Fighters] is rather congested" and it would "improve the general 'view'"[87] to redesign the gunsight.[88] He proposed removing the reflector screen entirely and using the cockpit windscreen instead as the reflective surface for the gunsight projection. He wrote, "As all of these planes [Single-seater Spitfire Fighters] have B.P. [bulletproof] screens the question of flexibility does not arise. Sight production [e.g. gunsight production] will not be interfered with as the existing optical unit can be used as the projector and the B.P. screen as the reflector."[89] To accomplish this task, researchers at R.A.E. Farnborough embarked on an ambitious development program that lasted through the end of the war.

The reasons for the removal of the reflector screen directly relate to the idea of the See-Through Graphical Interface as a device to project graphical information directly into the pilots' view of their material environment. Turning the windscreen itself into a reflector screen would transform the entire cockpit into a transparent interface for the computing unit. Likewise, the windscreen would become the reflector screen, enlarging the area on which to project the sighting graticule and eliminating awkward and inefficient sighting mechanisms of earlier designs.

This would build upon the visual advantages noted in the transition from the peephole to the reflector sight; a pilot would no longer have to look through a constricting device in order to target an enemy plane.

The proposal to use the windscreen as the reflector screen was enthusiastically received. The Wing Commander in charge of the trials for a prototype of the new gunsight wrote:

The opinion of pilots of the Unit and others who have flown the aircraft is that the forward view is very much improved and psychologically the installation is good because the pilot does not have to concentrate on the sight so much as previously, as he is not conscious of looking through a reflector glass.[90]

No phrase better expresses the forces that led to the development of the STGI than the test pilot's observation that he was "not conscious of looking through a reflector glass" when using the windscreen gunsight. Not satisfied with the larger window utilized by the reflector screen, researchers desired to further enlarge the aperture size of the interface motivated by the desire to use the entire cockpit windscreen as a surface of projection. To not be conscious of looking through a reflector glass is a sought after attribute of the design of the STGI. If realized, pilots are cognizant only of their material environment and the information superimposed onto it. The interface literally disappears from view, an attribute still pursued today in the design of STGIs: now, Head Mounted Displays can track users' head movements enabling information to be accessible from a full 360 degrees.

Despite the Chief Superintendent's enthusiasm for the project, because of numerous technical hurdles, it did not prove easy to develop a prototype suitable for mass production. In order to be effective as reflector screens for the gunsight, the bulletproof windscreens had to be manufactured to such a high level of precision that they were impossible to produce in quantities needed for the war effort. In addition, the light bulbs used in backlighting the aiming mark were not of a sufficient brightness to make the gunsight useful for daytime flying.[91]

FROM GUN-SIGHTS TO HEAD MOUNTED DISPLAYS

With the war's end there was a diminished sense of urgency surrounding the development of new gunsight technologies, but the British military continued to invest considerable resources into Head-up Display research and other devices for the visual presentation of flight and targeting information in aircraft cockpits. As should now be evident given the history of the gunsight provided in this chapter, R.A.E. Farnborough was pivotal to the development of the HUD both in the decades following World War II and the design of predecessor technologies, such as the Gyro Gunsight. J.M. Naish's work at the facility, discussed earlier in this chapter, was an extension of research at the facility that had already been full bore in the service of the war effort.

This chapter has demonstrated how a new category of media interface devices facilitates the identification of computing technologies going back to the first half of the twentieth century. The Mark II Gyro Gunsight was the first STGI, for it was the earliest device to project graphical information into a pilot's line of sight as an aid for targeting moving enemy aircraft. In making this case, I focused on a few aspects of the Mark II's design: the transition from a peephole sight to a transparent reflector screen and subsequent enhancements to enlarge the screen. Both were characterized by efforts to make the interface device itself disappear, so that graphical information could be presented to pilots in such a way as to minimize the obstructions to their field of view and improve pilots' capacity to act on this information.

The design of this first STGI was intended to enhance knowledge over one's material environment, a form of visual augmentation that links the Mark II gunsight to subsequent STGIs, such as the HUD and AR-HMD. From the first STGI to today's Land Warrior program, the military has shown an abiding interest in technologies that enhance the capacity to act in one's material environment. One can imagine what the future holds for this technology. When the military develops successful technologies, consumer applications are likely to follow. AR-based devices and software are poised to transform consumer interaction with computing devices, for there are a growing number of AR applications available for the iPhone and Android operating systems for a wide variety of entertainment and consumer purposes, including Google's recent announcement of Project Glass (see figure 4). The mobile foot soldier will become an AR-HMD enhanced consumer, hunting the most spectacular sales on the shop floor.

This historical study has far-reaching implications for the concept of the virtual as it has been defined in media and visual studies. The virtual has generally been linked to the opaque screen of media technologies. In subsuming the STGI to research on the opaque screen, researchers have oversimplified the relationship between the material and virtual in light of See-Through Interface devices. As suggested here, the STGI expands upon the history of the Graphical User Interface by identifying a class of interface technologies that have been mainly under-theorized. This history complicates a dualistic conception of the relationship between the virtual and the material, because the STGI imposes an information layer between users and their physical surrounding by projecting graphical information into their field of view. The insertion of this layer complicates any simple opposition between the material and the virtual, for the material itself becomes an integral dimension of the interaction.

Finally, the study presented in this chapter concerns a particular vision for the technology wedded to military research and development. It is, clearly, a deeply instrumental one, for with the STGI, the military aims to augment soldiers' lethality in the theater of combat by making their surroundings subject to a form of total information awareness. This vision for the technology is dangerous and limited, but it does not proscribe alternative uses for it. The next chapter considers critical engagements with the medium, such as artistic practices that utilize the interface in ways that trouble or directly contradict the military's deployment of the STGI. Instead of applications that extend the virtual to the material, the reverse will be explored: aesthetic practices that enable an encounter with the material in virtual and virtualized experience. Such practices could enable potent recombinations of materiality, the virtual and embodiment and, even, pose challenges to their current configuration in such a way as to encourage critical reflection on our shared cultural condition.

1
A soldier modeling the Land Warrior System.

2
A gyro gunsight in the cockpit of a Spitfire airplane. It is unclear what model this one is.

3
This diagram depicts the optical system of the GGS. According to Clarke, "The lamp shines through the graticule (G) onto the gyro mirror, which reflects the image of the graticule onto the fixed mirror (M) which reflects through the lens (L) onto the pilot's reflector screen" (Clarke, 1994: 171). Diagram adapted by author.

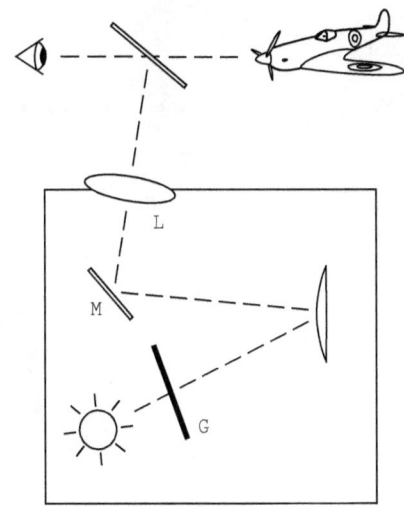

4
Announced in April 2012, Google Glass™ is Google's foray into Augmented Reality Head Mounted Display technology. The AR-HMD allows one to navigate Google's suite of services, including Google Maps, Shopping and Chat.

Encounters with the Material: Krzysztof Wodiczko and Site-Specific Media Art

Since 1981, Krzysztof Wodiczko has been crafting site-specific outdoor projections, artworks that have drawn attention to a viewer's surroundings in order to provide a critique of the public function of often overlooked monumental architecture.[92] His work addresses these overlooked structures in hopes of compelling viewers to connect their relevance to ongoing social and political problems. As distinct from the military uses and applications put to See-Through Graphical Interface technology discussed in the last chapter, Wodiczko's public artwork is a prescient encounter with technologies to visualize information away from the immobile screens of desktop computers and televisions. While See-Through Graphical type technologies are already being mined for their military, entertainment and commercial potential, Wodiczko's projects, by contrast, are distinctive for their critical address of the experience of our built environment. Building off of research on site specificity in art, in this chapter I argue that his media work is a cultural practice that opens an investigation of three interrelated aspects of site: its social and political meaning, its disciplinary function, and its materiality. By opening an encounter with the materiality of site, work like Wodiczko's is a critical address of the virtual in our current cultural condition. This analysis challenges how mobile technologies, like STGIs, are characterized today, as technologies that speak to the heightened expansion of the virtual into the experience of the physical environment.

Krzysztof Wodiczko has been making public art since the 1960s. His work has explored many different forms, spanning art objects and interventionist public performance to image and video projections. Since the 1980s, Wodiczko has become widely known for his image and video projections onto public monuments and the facades of urban architecture. Using high-powered projectors, these projections consist of imagery that examines or reveals the social, cultural, and/or political significance of site. In his initial work, these projections consisted mainly of static imagery, but his more recent work makes use of moving images.

In one of the more provocative readings of Wodiczko's work, Dick Hebdige suggests that Wodiczko seeks to cultivate a feeling of the uncanny as a political tactic in his artwork.[93] The uncanny arises out of his art objects, because he takes everyday objects and transforms them into something altered, unexpected, haunted. With his *Homeless Vehicle*, for example, Wodiczko turns a shopping cart with its familiar use by the homeless as a mode of transport and habitation into a striking militarized vessel "as if transplanted from another dimension."[94] Hebdige quotes Patrick Wright's characterization of it as "unlike anything that has ever existed before, and yet deliberately engineered out of resemblances to things familiar."[95]

Though he diverges from Freud's celebrated examination of the term, Hebdige draws on Freud's concept of the *unheimlich*, or the uncanny, in his study of Wodiczko's art. For Freud, the uncanny object "proceeds from something familiar which has been repressed."[96] The repressed in Wodiczko's work is the social antagonisms and conflict surrounding the original object, what Wodiczko refers to as the "problem issues"[97] that are of topical concern in his artwork. With its reimagining of the everyday shopping cart, the *Homeless Vehicle* casts an indirect light on the rampant homelessness and social displacement Wodiczko observed in his visits to the United States.[98] The raw materials for his projects are those overlooked everyday objects that are open secrets about the failings of society.

The *unheimlich* is key to understanding Wodiczko's site-specific projections as well. For Freud, places and built environments can also be *unheimlich*. The *unheimlich* derives from the German word, *heimat*, which refers to one's hometown or birthplace and the feeling of closeness and connection to it. Thus, a common definition of the *unheimlich* is the experience of one's hometown as a haunted, unfamiliar place. Freud quotes the Grimms' *German Dictionary* about the close connection between the *unheimlich* and *heimat*, "At times I feel like a man who walks in the night and believes in ghosts; every corner is *heimlich* and full of terrors for him."[99] In contrast to his art objects, Wodiczko's projections produce the feeling of the *unheimlich* in their capacity to alter one's perception of site. There is something undeniably creepy about the content of some of the projected imagery: skeletal hands holding instruments and machine guns, eyes without faces, and hands without arms or bodies that grasp handguns abound in his work (see Figure 1). But what contributes to their spectral uncanniness is that the site becomes bathed in the light of these ghostly projections; the site becomes haunted by the images. As ghosts are liminal beings, neither dead nor alive, neither physical nor fully disappeared, so too are projections, which are neither material nor fully immaterial, insofar as we are able to see the light and experience them.[100] As I argue later, the projections' formal differences from the site (what Friedberg[101] calls their "liminal immateriality" in contrast to the site's materiality) heightens the projections' uncanniness.

If the uncanny is a return of the repressed, what is the repressed of site? In other words, what is revealed about site in Wodiczko's projections? In order to answer these questions, I move away from the literature on the uncanny in cinema theory and focus on the role of place in site-specific media work, because what is distinctive about such work is that it occurs in public settings and not the darkened rooms of theaters. I link the "problem issues" Wodiczko identifies of topical concern in his work to Miwon Kwon's conceptualization of interrogative site-specific installation art, or work that interrogates the sociopolitical conditions of site.[102] Kwon's work on site specificity has been influential, because she identifies site-specific art's break from earlier public artworks[103] that did little to respond to the sociocultural conditions of site. In distinguishing between site-specific and site-generic art in terms of how responsive the artwork is to the conditions of site, Kwon insists that the site-specific work only makes sense in and is appropriate for the particular site for which it was crafted.[104]

If we apply Kwon's conceptual approach to Wodiczko's work, we can understand Wodiczko's projections as one of the earliest examples of interrogative site-specific media art. But because Kwon writes primarily about pieces that originate out of the plastic arts, there are distinct limitations in her approach to the analysis of the formal characteristics of media work. Therefore, it is necessary to go beyond Kwon's conceptualization of site specificity and recognize that Wodiczko's projections not only address the sociocultural conditions of site, but are responsive to the site in the form of a critique of the disciplinary function of built environment and an indictment of our oblivion to its social and political history. What is distinctive about Wodiczko's art is that these multiple ways of engaging with site are marshaled in the creation of an encounter with materiality, an encounter that is foregrounded by the immaterial projection of the sociopolitical conditions of site. In order to develop this claim, the following two sections investigate how Wodiczko's work engages with the sociopolitical conditions of site and interrogates the built environment as a social formation that exerts disciplinary power over inhabitants.

ENGAGEMENT WITH THE SOCIOPOLITICAL CONDITIONS OF SITE

Wodiczko refers to his own work as a practice of interrogative design, which Rosalyn Deutsche has observed consists of "first, a symbolic operation by which concepts are visualized as external realities and, second, a rhetorical device for speaking with clarity at a distance."[105] In his later work the "concepts" that are visualized refer to hot button "problem issues"[106] of concern to Wodiczko, such as homelessness, war and militarization, migrant labor, and others. Critics have mainly focused on these problem issues and how his work critically engages with them (Hebdige on homelessness;[107] Deutsche on redevelopment and gentrification;[108] Smith on spatial differentiation[109]), for the visual language that he develops for the projections' overt content derives from reflection on these issues. In his projection on the *Arco de la Victoria* (Victory Arch) in Madrid, Spain (1991), for example, Wodiczko produced visual imagery critical of Spain's military involvement in the Gulf War. On the supporting columns of the arch, two skeletal hands were projected: the right one, holding a machine gun; the left, a red gas pump. Above the skeletal hands on the arch's attic was projected the question "*¿CUANTOS?*," in reference to the cost of war in terms of the numbers of dead, the price of oil, the cost of war, etc.

The *Arco de la Victoria* piece, like much of his projection work, is responsive to the site in that the location is both the inspiration for the artwork's content and is used as the surface for projections. The content interrogates the arch's function as a memorial to war, while simultaneously recalling the bloody legacy of Spanish militarism and focusing attention on the wars in which Spain continues to participate. This linkage between site and content is integral to Wodiczko's art, because he seeks to address the unconscious power of forgotten monuments:

Franco, who ordered the monument to be built, is petrified within it but people pretend not to see him. I don't want to bring him back to life—that doesn't interest me. What interests me is seeing the connection between this monument and present day concerns.[110]

Despite their scale and imposing presence, public monuments, like the *Arco de la Victoria*, are "neglected" or "overlooked,"[111] because they disappear into the background and are no longer seen. Wodiczko is fascinated by how their power accrues in their disappearance. Monuments represent norms and values that are preserved in the public's mind despite a sense that society might have superseded them. His *Arco de la Victoria* projections link historical aggressions to Spain's continuing participation in wars, opening up reflection on how war and militarism continue to inform the social imagination.

Put another way, Wodiczko's work interrogates the overlooked sociopolitical conditions of site. When understood in this way, Wodiczko's work fits Miwon Kwon's "paradigm" of interrogative site-specific art,[112] where site "is imagined as a social and political construct," and the function of the artwork is "a proactive interrogation," which "manifest[s] a judgment (presumably negative)—about the site's sociopolitical conditions."[113] The interrogative public artist's relationship to the site is not complementary, but "antagonistic." Kwon analyzes Richard Serra's *Tilted Arc* (1981-1989) as an exemplar of this type of public art (see Figure 2). The site, Federal Plaza in downtown New York City, was heavily used by pedestrians and people on work breaks. Serra's piece was a purposive interruption of the space, a massive 12-foot-high, 120-foot-long metal sculpture that effectively cut the plaza in two. Because of the disturbance it created and its perceived lack of aesthetic and social value by a vocal public, the installation became mired in "rancorous and vehement" controversy, which ultimately led to its early removal in March 15, 1989 after five years of public hearings, lawsuits, and media coverage.[114] Serra's piece was interrogative, because it was a critique of the sociopolitical conditions of public space, like parks and plazas, which are imagined as a "coherent spatial totality." According to Kwon, *Tilted Arc* "literalized the social divisions, exclusions, and fragmentation that manicured and aesthetically tamed public spaces generally disguise."[115]

In light of Kwon's account, Wodiczko's projections can also be characterized as interrogative in that they highlight the sociopolitical conditions of public monuments and architecture. Jeffrey Skoller has similarly noted that avant-garde filmmaking has been preoccupied with memorializing the sociocultural history of place through cinematic reflections on the present.[116] Whereas the films that Skoller discusses engage this through dynamic montage of contemporary and historical film footage, the *Arco de la Victoria* projection breaks from cinematic form by drawing attention to overlooked structures themselves by overlaying imagery directly onto site. By projecting his work onto war monuments, which are intended to memorialize past wars and the dead who were lost in them, Wodiczko's work implies Spain's continuing possession by militarism. This critique of shrouded, hidden values, like militarism, is characteristic of Wodiczko's pieces, which expose the conditions of the site and suggest how these structures speak to values still enshrined in the present.

Kwon reads the development of interrogative site-specific art primarily in historical and institutional terms. Its formation was a reaction to the prevailing "assimilationist" tendency in public art, one that gravitated towards "accessibility" and "usefulness" in the decades before the 1980s.[117] Serra challenged the assumptions of this assimilative paradigm in public art and posed a counter-model of site specificity. Though Kwon's is a useful historical periodization, there are distinct limitations in locating the interrogative function of site-specific work solely in the way it responds to or reacts against paradigms internal to the field of public art. I propose two alternate ways to read Wodiczko's work: as a critical address of the disciplinary function of site and an investigation of materiality. Both move beyond Kwon's historical periodization and suggest ways in which his interrogations are more provocative than their critique of the sociopolitical alone and open up ways to further understand their distinctiveness as a formal exploration of materiality.

ENGAGEMENT WITH SITE

If the content of Wodiczko's projections reveals the sociopolitical conditions associated with site, the location and placement of the work address the site's role in enforcing and reproducing social norms in society. His work interrogates the built environment, specifically its urban design, as a social formation that exerts disciplinary power over inhabitants. About this power, Wodiczko writes:

What is implicit about the building must be exposed as explicit; the myth must be visually concretized and unmasked. The absent-minded, hypnotic relation with architecture must be challenged by a conscious and critical public discourse in front of the building.[118]

Wodiczko draws on Foucault's influential studies of the history of modern disciplinary institutions in defining this exercise of state power in the construction of space and the implantation of this power in individuals. "As Michel Foucault taught us, extensive experience and intensive disciplining processes have, since the eighteenth century, transformed our bodies from rural to urban, from undisciplined to disciplined... Architecture was a didactic mission."[119] Here, Wodiczko is paraphrasing Foucault's well-known analysis of the birth of modern discipline, in which he recounts the origin of Jeremy Bentham's panopticon in the late eighteenth century as a physical structure that both incarcerated the bodies of inmates and served as a potent symbol of the modern state's consolidation of power to discipline individuals. Foucault's observations about the birth of modern disciplinary institutions are echoed in Wodiczko's claim that, "Our position in society is structured through bodily experience with architecture."[120] War monuments, banks, museums—the design of urban space—are more than simply concrete forms; they are fetishistic reminders of the unconscious forces that regulate behaviors. A public structure, like a war monument, speaks to the legacy of militarism, but it also has a role of implanting these values as norms within us, because the design of our built environment is part of the symbolic universe that interpellates individuals as subjects.[121]

What is counterintuitive about this power is that it is exercised in those places we feel most comfortable. Familiar spaces are where a maximum amount of discipline is in evidence. The effective hegemonic functioning of public monumental architecture is precisely in its capability to go unnoticed. Referring to Guy Debord's work on psychogeography, Wodiczko characterizes urbanism (which includes city planning, public architecture, urban design, etc.) as "the material foundation" for those "symbolic, psychopolitical and economic forces" which interpellates individuals.[122] Our habitual movements through space are evidence of the existence of these forces over our seemingly autonomous decisions. We have internalized the state's norms, and this is illustrated by the fact that we do not balk at the war monuments or the ornate churches we pass daily.

Wodiczko's projections are interventions into this daily remonstration of disciplinary force. Recalling Hebdige's characterization of Wodiczko's work as opening the door to the *unheimlich*, his projections trouble the familiarity of home. By dressing the familiar in foreign projections, it is changed; it becomes unfamiliar. Home is made *unheimlich*, because it is dispossessed of its cozy familiarity. For those who are receptive, this is a haunting that compels one to look closer and scrutinize one's built environment, especially those details that are generally unseen. One might notice, for example, the triumphal arch under which one walks on a daily basis and appreciate details, like the impressiveness of scale or the complexity of ornamentation, but structures become *unheimlich* (one "actually sees them"[123]) when one recognizes their disciplinary function within the constellation of forces that converge on an individual in the experience of site. It is in these moments when we "expose" site's function as "explicit," "unmask" its "myth,"[124] and engage in critical discourse about it. Wodiczko's work on and with site, in other words, is a traumatic resurfacing of its repressed.

From this perspective, it is necessary to deepen Miwon Kwon's claim that interrogative site-specific art engages its sociopolitical conditions. Her account does not sufficiently account for how artwork might be more sophisticated in its interrogation of site beyond an identification of explicit social or political issues found within it. Wodiczko's work expresses a concern with the exercise of social power and disciplinary force that converge at the focal point of site. What substantiates this claim is that despite the fact that the projections' content has shifted among different problem issues over the years (e.g. militarism, migrant labor, homelessness, etc.), his work continues to be situated in sites that have overt disciplinary significance. This is an indirect acknowledgement of the difficulties in challenging sociopolitical conditions in face of the disciplinary function of site. Critical discourse is blocked when we are unable to recognize how the built environment enforces and reproduces the social norms against which artwork struggles.

The shifting subject matter of Wodiczko's work also suggests the ephemerality of interventions based in video projection. In contrast to the materiality of site, video projection is immaterial, temporal and temporary. In other words, site is not only sedimented social power, but its material form is the surface for projections. The next section considers at length the formal qualities of Wodiczko's projections, where I draw on conceptual frameworks familiar to new media studies and argue that his projections onto site are an investigation of site's materiality. A point of entry into this aspect of his work is the recognition that his projections seem a prescient investigation of the possibilities of recent digital technologies, such as Augmented Reality. Wodiczko's works are an exploration of the aesthetic possibilities in the superimposition of virtual images onto one's built environment, a kind of informational interface with one's physical surroundings that is becoming widespread with the popularization of mobile computing devices.

ENGAGEMENT WITH MATERIALITY

Let's first flash forward to research on contemporary media technologies in which there is growing interest in alternatives to our interaction with data beyond the opaque window of the computer screen. Of particular interest are mixed reality interfaces, like Augmented Reality, that have the potential to reconfigure our experience of the built environment. In one of the more sharp-minded theorizations of this kind of interface, Jason Farman has insisted that mobile computing enables a "collaboration of material and digital spaces."[125] One's geographical location, for example, is increasingly integrated into our interactions with mobile technologies. This is akin to Giannachi and Kaye's insistence on the growing hybridization of virtual and physical presence with mixed reality media usage.[126]

As an illustration of this collaboration, Farman examines Streetmuseum, an Augmented Reality app for the iPhone released by the Museum of London in 2010.[127] Streetmuseum enables users to view historical photographs and paintings in the museum's collection superimposed on their actual location. When standing outside Buckingham Palace, for example, one is able to view a picture of British suffragettes being arrested for protesting. Looking through a mobile device at the specific location, it is as if the image were projected onto the site itself.[128]

I address differences between them towards the end of this section, but with regard to the experiences each provides, there are points of comparison between Augmented Reality applications, like Streetmuseum, and Wodiczko's site-specific projections. Both superimpose visual imagery into the view of one's surroundings, and the content is dependent upon the viewer's physical location. Additionally, in both cases the imagery is not only intended to draw attention to its content, but also to the built environment.[129] The overlaying of images invites viewers to discover something about the site that is not apparent to the naked eye, such as some detail about the social or political significance of the location, prompted through curated visual content that illuminates the site's meaning.

The conclusion Farman draws about the role of mobile computing in mediating site, however, is different from those drawn about Wodiczko's outdoor projections in previous sections. Farman recognizes in such technologies the growing interpenetration of material space by an informational or virtual dimension. He writes:

The move from personal computing to pervasive computing, a shift characterized by the move from immobility to mobility, has allowed for online space to interact with material space in unprecedented ways.[130]

Farman identifies a substantive erosion of the distinction between the material and the informational with the rise of mobile computing. One's physical surroundings become an "information interface"[131] because a person's geographical location is an indexical marker to online information accessible by mobile computing devices. Farman writes of the "collaboration" of the material and the informational, but the focus is on the ways in which physical space is being subsumed to informational processes.

Farman's reading does not consider with enough complexity our relationship to the material, because it draws heavily from existing concepts of the virtual in new media studies. It has been argued that digital technologies have extended and enhanced the possibility of virtual-type experiences.

While it has been productive for the study of various developments in mediated culture, the concept of the virtual has led to an impoverished idea of media's role in connecting us to our physical surroundings. In new media research, the material has primarily been conceived in its disappearance. By applying insights derived from the analysis of Wodiczko's site-specific media projection, we can discover resources for an alternative account to the material that does not presume its disappearance.

Chapter 2 reviewed research on the virtual and documented how the concept has been instrumental in revealing transformations brought about by media and informational processes, but this research has relied on a systematic exclusion of a concept of the material. Examples of this proliferate in the literature: from Baudrillard and Derrida's early insistence that the virtual has come to determine cultural processes to more recent work on disembodiment in online environments. The virtual has enabled researchers to identify the myriad ways in which social, economic and cultural phenomena are being dematerialized with the advent of informational processes. This has constrained inquiry by providing a singular interpretation of these developments. The material has either been completely excluded, or acknowledged only in its demise. In other words, the virtual has been conceived of as the conceptual obverse of the material with the material perceptible only in its disappearance. In this regard, the material is a kind of "mute facticity"[132] shaped or acted upon by the virtual and has not been given consideration as a constitutive element of interactions with media technologies.[133]

When considered in this way, it should come with little surprise that Farman depicts the material as subject to colonization by informational processes. The hybridization of the material and the informational results in the transformation of physical environment into an "information landscape,"[134] akin to existing accounts of the precedence of the virtual. What's needed, however, is an analysis that recognizes with more complexity our interaction with the material, and Wodiczko's work can help us refine our thinking. Before developing this account, it is important to acknowledge differences between Wodiczko's projections and AR-enabled interfaces, like Streetmuseum. Whereas Streetmuseum is available on Internet-capable mobile computing devices, Wodiczko uses high-powered 35mm slide projectors, only beginning to use video projectors in the 1990s. Because of the nature of the projections, Wodiczko's work is staged as public events with large numbers of people in attendance. With AR, on the other hand, the content is available to the public (to be more precise, only those who have smartphones), but experienced privately on a handheld device. While Streetmuseum draws on historical photographs and paintings from the museum's collection, Wodiczko stages scenes to illustrate the conditions of site.

Despite these technical differences, when we consider the formal characteristics of both, similarities emerge. What is striking is their shared exploration of the formal differences between the materiality of site and the immateriality of the projected content. Both require the viewer to venture out into the built environment in order to see the imagery projected onto site. The contrast between the light and impermanence of the projected images and the site's materiality—the surface for the projections—is a significant part of what makes the experience compelling. In order to characterize this dimension of the work, let's reconsider Anne Friedberg's definition of the virtual, introduced briefly in the first half of this chapter:

Any representation or appearance (whether optically, technologically or artisanally produced) that appears 'functionally or effectively but not formally' of the same materiality as what it represents. Virtual images have a materiality and reality but of a different kind, a second-order materiality, liminally immaterial.[135]

As a metaphor substitutes one meaning with another, a projected image stands in for the thing itself. Though this comparison is "functionally" accurate, a difference between a metaphor and a projected image is that a projection also differs formally from the thing it represents. The projected image is liminally immaterial, made of light and darkness.

One way to conceptualize the history of the media is as a record of trying to cover over this formal difference between the liminally immaterial projected image and the thing it represents, to create an experience in which the virtual image is mistaken for the material object. In her discussion of the various technological and perceptual developments around the virtual image, Friedberg reflects on the architectural details of the movie theater and how historical changes to its design were made to enhance its capacity for enhancing the illusion of cinema.[136] The construction of movie theaters facilitates a viewer's willing suspension of disbelief in the virtuality of the images. With their darkened interiors and prosceniums that suggest theatrical stages, movie houses are crafted so that spectators can temporarily lose themselves in the world on display. Spectators suspend their recognition of the virtuality of the images and enter them in imaginative construction as real places.

The site specificity of media projection can upset expectations about the suspension of the material during media spectatorship. The settings in which viewers experience Wodiczko's projections are not cinema's silver screens and darkened theaters. Wodiczko's projections take place in public locations in order to address the site and its sociopolitical conditions. He projects onto the built environment, and the overlaying of virtual images onto site highlights the formal difference in the materialities of site and image. (It is as if the theater itself is illuminated and the screen left intentionally dark.) For uninformed viewers, the realization that the projections' sociopolitical content refers to the structures under illumination can come as a surprise. (One experiences his work when attention is arrested by the uncanniness of the projections; when one's movement through space is altered.) This uncanniness draws attention to the buildings, the monuments, and the built environment serving as the screen for projections. The virtual images float spectrally on the buildings' surfaces, which is a solid materiality. Under illumination, the material is brought forth as that which exists beneath the light of projections. In this way, Wodiczko's work creates awareness of the formal difference between the projected images and the site, the virtual and the material. Wodiczko's work is an encounter with materiality, one that is foregrounded by the immaterial projection of the sociopolitical conditions of site.

What is not being suggested is that the material gives meaning to site, or is a locus of contact or interaction that is a more authentic relation to one's built environment. In both of these senses, the material is imagined as a kind of ground or truth from which meaning derives. Critical of a similar tendency, Schopenhauer once wrote that "everything objective, extended, active, and hence everything material" is regarded "as so solid a basis for its explanations." This skepticism with materialist explanation and the diverse forms it can take is illustrative of the confluence of complex structures of meaning, histories and cultural processes that attend our understanding of the material. Like the virtual, the material is constructed in interweaving systems of signification, which fashions the manner in which we evaluate, understand, sentimentalize and relate to our built environment. Though the theorization of the concept of the virtual has led to a refined understanding of how the media are integral to our interactions with the world, it has simultaneously obscured a concept of the material and the different modalities of contact with it.

RUPTURE OF THE VIRTUAL: TOWARDS MEDIA PRACTICES OF MATERIALITY

I want to open up the implications of this reading of site-specific media art by expanding upon the conclusion drawn at the end of the last section: that Wodiczko's projections are an encounter with materiality, one that is foregrounded by the immaterial projection of the sociopolitical conditions of site. Three linked speculative claims about the implications of this conclusion for critical media practices are explored in this section. The first claim applies this notion of an encounter to a range of cultural practices, not just Wodiczko's projections. His work is a powerful example situated in media practice, but this encounter is not exclusive to his work alone. To the extent that a range of cultural practices share this preoccupation with the material, I argue that such encounters are responses to the growing experience of the virtual as a cultural condition, what I earlier referred to as the mounting immaterialization of social, economic and cultural phenomena because of informational processes. When contextualized in a way that enables us to link the work to broader concerns, these encounters can produce a critical understanding of our condition relative to the virtual. Finally, I develop a brief typology that organizes cultural practices according to their capacity for a critique of the virtual and define a type of practices that "rupture the virtual."

In the last section I argued against the view that AR is simply a technological expansion of virtual processes, a transformation of the built environment into an "information landscape." This position results from a concept of the virtual in which media technologies are seen to enhance or expand the reach of virtual processes. Despite my insistence on the conceptual limitations in this application of the concept to new media, this should not be confused with a rebuke of the concept's usefulness in understanding and characterizing contemporary culture. Indeed, many of the premises found in the theorization of the virtual are compelling, especially the underlying claim that our cultural condition is one increasingly subject to forces of virtualization—the argument that most, if not all, experiences are inflected by the virtual to some degree. This is evidenced by a reliance on the media to extend one's sensory perception of the world and the dematerialization of forms of social interaction and work.

Wodiczko's art is situated in the experience of the virtual in contemporary culture, and invites an interrogative address of it by creating an encounter with the material. The overlaying of projected images onto the built environment foregrounds the site's materiality, preventing the site from receding into darkness. Such an encounter with site's materiality challenges viewers' expectations about the circulation of images in a culture of heightened virtuality. This is to say that a recognition of the role of the material in Wodiczko's projections invites an interrogative address of the virtual. This recognition distances viewers from an embeddedness in the virtual and contributes to a critical position regarding it.

The sociopolitical content of Wodiczko's projections facilitates this encounter but is not necessary to it. The content is a reflection of Wodiczko's aesthetic choices and is not the only way to activate site. Content, in other words, is a tactic, not a key to unlocking the meaning of his chosen site. One can imagine a range of practices that produce a recognition of the material with a range of content, including—but not limited to—the construction of site-specific content that relies on fictional narratives concerning material objects in the site, and the stripping away of sociopolitical content in order to foreground formal differences between the virtual and material. This invites analysis of a range of contemporary media and performance artists, who have explored the material in their site-specific work. Chapter 5 examines work by artists including Janet Cardiff, George Bures Miller, MAW, Daniel Dean and Ben Moren. Media artists, however, are not alone in their preoccupation with this form of critique; it is possible to expand the horizon of this research to include non-art based cultural practices. The conclusion of this book reflects upon examples of this. Site-specific media projections during the Occupy Wall Street protests, for example, helped to illuminate the insight that space can be more fluid, contestable, or transiently occupied, despite the disciplinary pressures exercised upon the proscribed use of site.

In this regard, cultural practices that aim to refine their content into a direct interrogative address of the virtual are compelling, precisely because they have the potential to reveal in less overdetermined ways the viewer's relationship to the material. Though penned in a different context, Theodor Adorno's essay "Lyric Poetry and Society," about the poetry of the late nineteenth and early twentieth century, can be a guide for understanding this type of work. Adorno resists interpretive positions that portray lyric poetry as "the most delicate, the most fragile thing" that seemingly rises above the "bustle and commotion of the everyday."[137] Instead, Adorno offers an opposing conclusion about lyric poetry. He contends that it is precisely deadlocked social antagonisms that constitute lyric poetry as the most delicate, the most fragile thing. Because of this, analyses that take into account social context are of the utmost importance. When Adorno's insight is applied to cultural practices that engage the material, we can conclude that the more practices are able to open encounters with the material without veiling sociopolitical content, the more sharply defined their critique of our cultural condition. I contend that they "rupture the virtual," because of their more realized critique of our cultural condition relative to the virtual.

1
Krzysztof Wodiczko, *Arco de la Victoria, Madrid*, 1991. © Krzysztof Wodiczko. Courtesy Galerie Lelong, New York

2
Richard Serra, *Tilted Arc*, 1981. Image: © G. Lutz.

Materiality in Site Responsive Media Art

Krzysztof Wodiczko's projection work established an early and influential approach to site-specific public media art. Since the 1980s, there has been a rapid expansion in the number, variety and technical sophistication of outdoor media art. Today there are a number of artists who engage in projection-type work but operate with different models for site specificity. This chapter investigates three groups of media artists and collectives who create work that make use of site in innovative ways, but in configurations differing from the examples previously examined. In contrast to the previous chapter's focus on site specificity in media art, I employ the term site responsive to characterize the works discussed in this chapter. Site responsive is intended as a more inclusive notion of how media art responds to location. While site-specific art exhibits the characteristics that were discussed in Chapter 4,[138] site responsive media art, by contrast, may exhibit some (but not all) of these characteristics. Site responsive, in other words, is used here as a more open-ended category for media art that includes both site-specific art and work that is agnostic about site specificity.

An investigation of the variety of site responsive models for media art will support the discussion of the material and the media's role in constituting knowledge about it. The works examined here include ones ones that critically investigate the material to those that blur or hybridize viewers' perception of their physical surroundings in such a way as to extend the virtual out into site. By examining projects that engage site in a multiplicity of ways, we can extend the utility of a concept of the material in the examination of media that frames a relationship to one's physical surroundings.

JANET CARDIFF AND GEORGE BURES MILLER: AUDIO AND VIDEO WALKS

We are all searching for the real, for the authentic experience... that we are actually on this earth ... We are building a simulated experience in the attempt to make people feel more connected to real life.[139]

Cardiff and Miller have been creating artwork since the early 1990s, and their projects span a wide variety of forms, from gallery installations and sonic environments to outdoor media pieces. I focus on a body of work they refer to as audio and video walks, because, in light of the site-specific work discussed in the last chapter, they make use of comparable sensory and media elements. The manner in which they have spoken about their intentions for the work, furthermore, resonates with the discussion of the material found in Wodiczko's projections. As suggested in the quotation above, Cardiff and Miller characterize their own work as an effort to "make people feel more connected to real life" or inspire a "renewal of a sense of place."[140] In an examination of a selection of their audio and video walks, I argue they can also produce an opposite effect: the extension of mediated experience out into the material, which blurs the boundary between the real and the invented. While Wodiczko's site-specific projections foreground the sociocultural conditions of site and its materiality, Cardiff and Miller's extend the virtual out into the participant's interactions with site. I begin with an examination of one of Cardiff and Miller's gallery based projects, which illustrates a preoccupation with the virtual and its relation to the material that is central to their audio and video walks.

Originally created for the Canadian Pavilion at the Venice Biennale in 2001, *Paradise Institute* (2001) consists of a plywood structure 17 x 36 x 10 feet with two doors that grant access to viewers (see Figure 1). When viewers enter, they find themselves on a theater's balcony with 16 seats. It is a model of an old-fashioned movie theater in extreme forced perspective and crafted in miniature to give the illusion that one is in a full-sized movie theater despite the sturcture's actual scale (see Figure 2). A movie is screened and participants listen to the soundtrack on headphones provided at each seat. Cardiff and Miller constructed the soundtrack in such a way as to confuse viewers about which sounds are part of the movie's storyline and which are recorded ambient sounds that are outside the movie's narrative, such as the sounds of coughing, chit-chat and rustling, a ringing cell phone, and a woman who whispers intimately into one's ear. In interviews, Cardiff and Miller have said that there are, in fact, three discrete soundtracks to the piece. "One level gives you the experience of the film soundtrack. The second level is... a 3D soundtrack of an audience around you. The third is a… sound environment of scenes that unfold around you in the darkness."[141]

The audience strains to distinguish between the movie's soundtrack and external noises, real and fake audience noises, between sounds occurring inside or outside the structure. The piece unsettles conventions surrounding spectatorship in theatrical settings, which makes the experience discomforting. The audience's attention jitters between the film's soundtrack and the recorded audience noises. This shuffling of attention thwarts one's ability to settle into the immersiveness of a cinematic experience. *Paradise Institute* is less about the narrative content of the film being screened than this struggle to comprehend the multiple soundtracks. In other words, the piece examines the audience's positionality as spectators in the constructed environment of the theater by unsettling the expectations that normally govern it.

In Jonathan Crary's analysis of modern spectatorship, he charts how our capacity for attention is the outcome of over a century and a half of construction.[142] That we can sit in fixed attention for the duration of a Hollywood movie hardly seems a feat of disciplined self-control, but when understood alongside myriad ways in which focused attentiveness is demanded in contemporary forms of consumption and labor, such as full-time computer work and assembly line manufacturing, the importance of the construction of attentive subjects becomes clear.[143] As part of his analysis of the modern attention economy,

Crary also notes the rise of psychological disorders associated with an inability to properly embody attentiveness, including forms of schizophrenia and Attention Deficit Disorder.[144] Symptoms associated with these disorders can include hyperactivity, disorientation, confusion, inability to concentrate or focus, and others. These are precisely the kinds of effects that Cardiff and Miller produce in *Paradise Institute*. The disruptions to the narrative make it impossible to focus on the plotline, creating a disorientation akin to the disorders associated with inattentiveness. What an interloper whispers into your ear, for example, becomes part of the unfolding sequence of events occurring on the screen. The screen does not define the entirety of the audience's experience of the piece, for the outside breaks in as a constitutive element.

This preoccupation with the inside and the outside, the constructed and the real, sets up a discussion of Cardiff and Miller's audio and video walks, which cultivate similar tactics. In contrast to their gallery-based work, the walks draw on a tradition of recorded driving and walking audio tours.[145] At the start of a walk, participants are given portable audio players with headphones (in later video walks people are given portable video players, such as digital video cameras or video iPods). Prerecorded narrators (frequently Cardiff herself) entreat listeners to follow their voice to various locations around the site. At stops along the way, listeners pause to look and listen to particular details about their surroundings. The walks are site-specific in that the stories are created for these precise locations, yet unlike walking tours, Cardiff and Miller's pieces tell a story about the place that is frequently fictional, but at times contain fragments of factual details.

To date, Cardiff and Miller have made over 25 audio and video walks and have developed a number of techniques to deepen a listener's involvement in the narrative. Given the large number of walks, it is difficult to generalize the techniques that they employ. I focus on a particular example, *Her Long Black Hair* (2004), an audio walk commissioned by the Public Art Fund for Central Park in New York City, to exemplify some of these techniques. The piece has the sensibility of a detective story, because listeners are given a packet of photographs at the start of the tour, which includes images of a mysterious woman with long black hair (see Figure 3). Cardiff's narratorial voice also has a conspiratorial tone to it, making it seem like we are in pursuit of someone or something. There is no discernible narrative structure to *Her Long Black Hair*, which consists mainly of the narrator's recollections and thoughts about various locations. In Cardiff and Miller's own words the piece is an immersive "investigation of location, time, sound, and physicality, interweaving stream-of-consciousness observations with fact and fiction, local history, opera and gospel music, and other atmospheric and cultural elements."[146]

Both *Her Long Black Hair* and *Paradise Institute* cultivate disorientation about what is happening inside and outside the structure of the work. While participating in Her

Long Black Hair, listeners hear gunshots in the distance, which the narrator claims, implausibly, is the sound of wild goats and pigs being slaughtered. (This claim is only slightly credible. There were pig farms on the land that eventually became Central Park, but it is doubtful that wild goats or pigs were ever shot in the park). Music drifts in—its origin unclear—the sound could be coming from within the park or it may be part of the recording. One of the photographs provided for the walk has the semblance of a historical snapshot of a crowd sitting in the park, further confusing participants about the walk's basis in fact or fiction. This continual shifting between the real and virtual, invented and factual, pre-recorded and live, is typical of Cardiff and Miller's audio and video walks.

Underscoring their recognition that the walks blur the real and the invented, Cardiff and Miller have variously referred to their audio and video walks as "physical cinema," "virtual reality," "3D narrative experience," and "film soundtracks for the real world." Though Cardiff and Miller refer to their work as virtual reality, it is a term they use knowingly as a metaphor, for none of it employs head mounted displays or computer-generated virtual environments. What they mean is that their work is a kind of immersive cinema in which our perception of reality comes to be experienced as mediated. As virtual reality substitutes the material with computer simulation, in Cardiff and Miller's work it becomes difficult to distinguish between what is occurring in one's physical environment and what is pre-recorded, simulated and virtual.

Despite the virtual reality of their audio and video walks, Cardiff and Miller have also claimed that they offer a "renewal of a sense of place"[147] or a sense of "connection" to a participant's environment. Cardiff and Miller have spoken about how their work is driven by an impulse to reconnect people to the physical world:

Walkmans have always been criticized for creating alienation, but when I first discovered the walking binaural technique I was attracted to the closeness of the sound and the audio bridge between the visual, physical world and my body. To me it was about connection rather than alienation. I think it's partially because the audio on the CD meshes with the audio in the "real" environment but it's also because sound does come into your unconscious more directly than visual information.[148]

Following up on this discussion of connection, Cardiff discusses a possible confusion between the invented and the real in the walks:

"There's a green door to your right, turn left at this corner, around the post. Watch out for the bike!" Some of what I say reinforces what you see in front of you so that my description parallels the voice inside your own head: "Yes there is a green door, yes there is a corner, I'll turn left." You are lulled into a complacency but then Bam! There is a disjunction: "Where is the bike?" I'm not sure why this technique interests me, but I think it works to push and pull people in and out of the experience.[149]

What Cardiff might be describing here is a version of the concept of salience, which is used to describe objects and ideas that take on heightened importance or fascination relative to other things in one's environment. When a narrator speaks knowingly of a "green door," for example, it gives listeners a sense of that object's heightened significance within the context of the story. (The next time a green door appears in the story, it is encountered with anticipation.) What is problematic about Cardiff's claim, then, is that a thing's salience primarily exists relative to the particular narrative in which it is told or contained. (All green doors do not take on heightened importance in one's life; only those in the story.) Because the walks are closed narratives, the objects and themes highlighted within them are salient only within the semiotic structure in which they are contained. Though Cardiff and Miller might wish that their walks produce a recognition or "renewal" of place and things outside of the walks' confines, the salience of objects exists only within the semiotic closure of a narrative and does not extend beyond it.

This semiotic closure is further evidenced in passing details found in some of Cardiff and Miller's walks that occur in public spaces outside the protective confines of parks, buildings or forests. In them the narrator reminds the listener to look both ways before crossing the street. (In a similar way she reminds listeners to close a door gently when on a library walk.) Participants are so immersed in the closed narrative of an alternate reality that they forget the rules and dangers by which this one operates. Cardiff and Miller's claim that they are interested in renewing a sense of place echoes Wodiczko's preoccupation with encountering the material. Instead of producing an awareness of site in ways that were discussed with regard to Wodiczko's site-specific work, Cardiff and Miller's audio walks make use of one's material surroundings as the site for the creation of an invented narrative. The audio walks are narrative structures that reference site and create salience surrounding particular objects within the work. The walks are thus site responsive, because they respond to and refer to one's surroundings, but they create a sense of disorientation about what is inside and outside, the invented and the real. To the extent that we can refer to Cardiff and Miller's media walks as virtual constructions, this disorientation is an effect of the blurring of the virtual with the material. The material is enfolded as salience into the closure of the virtual.

MINNEAPOLIS ART ON WHEELS

Since the early 2000's, a number of art groups have been exploring outdoor mobile projection as an artistic medium. Wodiczko's early projections were necessarily stationary, because they utilized large, high wattage slide or video projectors. Newer groups have made use of advances in projection technology, such as lightweight, lower wattage projectors that can be run off of batteries and deployed as mobile projection units. The rise of a DIY software programming culture has also led to the creation of tools to enhance forms of interactivity and immersion in outdoor projections. From virtual graffiti to bicycle-based mobile "happenings," there has been a proliferation in models for mobile outdoor projection, which provide additional insights into site specificity and site responsiveness considered here.

The next two studies are drawn from a burgeoning media art scene in Minneapolis, Minnesota, where I have worked since 2009. Minneapolis has become a hub of creativity around outdoor media arts. Steve Dietz, former curator of new media at the Walker Art Center and the founder of the SJ01 Biennial, has had a significant role in this. Since 2011, he has organized Northern Spark, a *nuit blanche* (dusk-to-dawn) outdoor art festival, which focuses predominantly on outdoor media art.[150] The two artist groups discussed next have both been commissioned by Northern Spark. Minneapolis Art on Wheels (MAW) and Mobile Experiential Cinema (MEC), which was founded by MAW members, explore the possibilities in outdoor media projection. Though both groups work with site-specific media projection, MAW has generally been less concerned about issues concerning site specificity while MEC continues to refine and explore new possibilities for immersive, mobile video projection in site responsive and site-specific ways.

MAW draws from a flexible group of collaborators.[151] Inspired by a visit to the Graffiti Research Lab in New York City, which had recently unveiled its Mobile Broadcast Unit (MBU), Ali Momeni founded the group in 2008. The MBU is a utility tricycle outfitted with a toolkit of technologies for mobile outdoor video projection; the basic setup includes a laptop, video projector, amplified speakers, battery power to provide electricity to all of this equipment (see Figure 4). The MBU became a platform for the creation of a wide range of outdoor projections, such as L.A.S.E.R. Tag, software that allows participants to draw on walls with swaths of projected color using only a laser pointer.

MAW's version of the MBU was built during a class on mobile projections at the University of Minnesota in 2008 (see Figure 5). They began "outings" with the unit that spring. Initially, they created the content using a mixture of digital video files and employed off-the-shelf tools, such as Flash. By the summer, Momeni had written a preliminary version of the software that came to define MAW's projection work, Livedraw. Originally written in MaxMSP (a newer version is written in C++), Livedraw is a software package that provides users with real-time video performance tools for mixing hand-drawings and real-time keying, layering and sequencing (see Figure 6). This allows users to create in real-time a wide range of projected content, including short video loops, hand drawings, and playback of prerecorded video.

MAW has organized hundreds of bicycle outings. Each one varies in content and duration, and as the following examples demonstrate, they are often guerilla in nature. The group does not seek permission to project on public walls, and because of this, outings generally take place in out-of-the-way locations (i.e. under bridges and on commercial buildings that are closed for the night) in order to avoid attention from the authorities. Despite the potential for run-ins with authorities, they have also organized outings in public, high traffic areas, where they can attract large audiences. As is the nature of MAW's work, the projectionists actively seek and encourage participation by pedestrians, soliciting short poems, drawings, etc. They have also asked the public to perform in the projections. This content gets composed into a collage-like video piece.

In conversations with Ali Momeni, he highlighted a handful of outings he thought were most successful. He recalled a series of projections on a building located at the intersection of 26th Street and Lyndale Avenue South, a busy stretch of the Uptown neighborhood in Minneapolis during 2008. The guerilla nature of these urban projections gave rise to unexpected and impactful encounters. What made this Uptown performance especially memorable for Momeni was the arrival and response of the Minneapolis police, which came in response to MAW's activities. In the ensuing conversation between the officers and the group, one member of MAW, using the MBU projected on a public wall, wrote the phrase, "I've lost everything, and I feel great." This prompted a conversation between the police officers in which one remarked how a friend had recently said something similar. The phrase became a disarming gesture that got the officers to become participants in the work and to recognize the value of MAW's public projections. The officers allowed MAW to continue their projections and left the scene.

A second memorable public projection, in October 2010, was more orchestrated. MAW has worked with institutions to create large-scale, public projections. "Seaworthy" was a multiple projection piece commissioned by the University of Minnesota for their Twin Cities campus (see Figure 7). Like the Lyndale projection, "Seaworthy" was participatory, consisting of wireless projections of audience members' faces, who were asked to close their eyes while reflecting on the prompt, "Think of the last time you were at the sea."

As both of these examples demonstrate, participation is central to the construction and experience of MAW's outdoor projections. Audience interaction is actively cultivated in the design of their work. What made these projects particularly successful was the public's unforeseen levels of surprise and interaction. For MAW, site is the location for social encounters and participatory interaction with a public that is drawn into the work. In this sense, MAW's engagement with mobile outdoor projection has more in common with relational art, what Nicolas Bourriaud has defined as "a set of artistic practices which take as their theoretical and practical point of departure the whole of human relations and their social context, rather than an independent and private space."[152] In MAW's work, the material becomes a screen for projections that facilitates active social interactions and participation.

MOBILE EXPERIENTIAL CINEMA

Daniel Dean and Ben Moren are both long-time contributors to MAW. Sharing an interest in experimental filmmaking, they formed Mobile Experiential Cinema (MEC) in 2011 for a Northern Spark commission to explore site-specific projection. They have subsequently collaborated on several works that explore comparable themes and ideologies. MEC makes use of the Mobile Broadcasting Unit developed by MAW as a platform for their own work, but Dean and Moren make significant changes to the outfit to facilitate their divergent approach to site-specific, responsive projection.

The last section highlighted how site was not a central concern in MAW's public projections. It is this relative disinterest that was one of the inspirations for the formation of MEC. Dean and Moren maintain a more consistent interest in site responsiveness and site specificity in their work. In an interview with Dean and Moren, Moren stated that in addition to a meaningful engagement with site, they were interested in examining three elements through their work: the urban environment, story and narrative elements, and live action sequences. In contrast to the earlier case studies, MEC's work explores the ways in which the virtual and the material intersect in media interactions. Site can be the screen for projection that extends the virtuality of a narrative cinematic experience. Their site responsive work explores the blurring or hybridization of the virtual and the real, presence and simulation, live and pre-recorded in media performance and art.[153]

Second Bridge is Wider, But Not Wide Enough is the title of Dean and Moren's first site-specific, mobile cinematic work. The narrative structure of the piece is organized around a small group of young adults who are discussing a missing friend. In the structure of the piece, the pretext for moving from site to site is to follow the group's recollections about their friend at the exact locations in which they occurred. At each site, the audience views a series of pre-recorded video shorts, which are projected directly onto the materiality of site. The opening sequence of *Second Bridge is Wider, But Not Wide Enough*, for example, occurs under the Hennepin Bridge, where the audience discovers the namesake for the piece on a memorial plaque (see Figure 8). A video sequence is projected directly onto the concrete bridge supports: a recording of the characters, in the exact same location, recounting their memories of their missing friend.

From the bridge, the audience is led by bicycle around Minneapolis to various locations where additional pieces of the story unfold. Even Minneapolis natives are unfamiliar with many of the sites. Dean said that their selection of sites was informed by an interest in "non-places" or "unprogrammed" locations. Moren added that the Situationist dérive was an inspiration for the identification of these non-places.[154] In a mix of technological savvy and ludic psychogeography, Dean and Moren first scouted sites by scouring Google Street View. Because Street View would often show buildings that were no longer standing or contain other digital mistakes, they would explore the areas in person in order to select specific sites that fit their criteria of unprogrammed non-spaces.[155]

The final sequence of *The Second Bridge* is one of the most memorable of the piece, because it contains the most intriguing sequences from the perspective of the site-specific filmic and narrative effect that they've created. While in front of an abandoned building close to the Mississippi River, the audience is shown a disorienting short video. The video is projected onto the building's facade, which has two staircases. Because the recorded video is of the same facade, when the video is projected onto the wall, it is misaligned with the actual staircases. This produces a sense of visual feedback, a visual disorientation that is accentuated in multiple physical and temporal dislocations that occur during the segment. A man, who might be the missing person who dominates the storyline, picks himself off of the ground in front of the projection wall. Is it a homeless person accidentally caught in the wrong place at the wrong time, or an actor planted there in advance? A cameraman steps out into the audience, and his footage is transmitted live into the video. This adds a layer of temporal dislocation to the piece, because it blurs the line between the pre-recorded and the live.

Dean and Moren expand upon their repertoire of techniques for site responsive cinema in a second project, *The Parade* (2012). Whereas *Second Bridge*'s had a loose narrative construction organized around a series of recollections, *The Parade*, by contrast, has a clear and conventional "Hollywood style" narrative structure.[156] The story concerns a kidnapping and attempted ransoming. Similar to the *Second Bridge*, there is a hybridization of the virtual and the material, the simulated and the real. The audience witnesses the kidnapping as a live-action sequence at the opening of the piece and over the course of the work becomes involved in an attempt to recover the kidnapped person by collecting a ransom and giving it to the kidnappers (see Figure 9).

Dean and Moren's work with MEC exists alongside other media artists who explore the relationship between the virtual and the real through their intentional blurring and hybridization. In an extended examination of British new media performance and the art group, Blast Theory, Benford and Giannachi note how the group's performances unsettle our distinctions between the virtual and real, live and the mediated, and the physical and simulated. In Blast Theory's performance piece, *Can You See Me Now?*, the group create a mixed reality game that blurs online play and real action. Groups of participants are divided into two, one seated at computer terminals and the other navigating the city using handheld computers with wireless network connections and GPS.[157] One group tries to track down the other who moves avatars through a virtual model of the same town. In light of Blast Theory's work and others like it, Benford and Giannachi argue that artists are making use of mobile and ubiquitous computing in order to "span physical environments and virtual worlds."[158]

When understood in this context, Dean and Moren are similarly engaged in the creation of work that hybridizes space, time and presence. But as a broad-stroke account of some of the transformations taking place with the use of mixed reality in performance and art, Benford and Giannachi's thesis about hybridization fails to capture some of the nuance in Dean and Moren's work. Dean and Moren's curation of non-places in their work, for example, suggests a fascination with physical environment in site responsive media that goes beyond hybridization. In choosing sites with which the audience has few associations, the environment becomes a "blank" site/screen for their work. This allows them to reimagine site as a screen for the projection of the virtual for the duration of the performance.

The hybridization of site is also in evidence in other details, especially in their use of Hollywood-style narrative in their second piece, *The Parade*. David Bordwell has described classical Hollywood style as an "aesthetic system" that includes a number of filmic and narrative conventions, including storyline development, cinematographic techniques, film length, color correction standards, etc.[159] As the dominant filmic convention, the Hollywood-style has led to the cultivation of what Bordwell, drawing on cognitive psychology, refers to as "schemata,"[160] or cognitive shorthands the brain uses to filter, process and interpret sensory information. These schemata facilitate ego-identification with a film's dramatic content. Schemata aid in imaginative integration into the piece, because viewers unconsciously rely upon them when forming impressions of a film. *The Parade* draws on these conventions to aid in the construction of the viewers' perception of site, especially those with which viewers have little or no familiarity.

In sum, *The Parade* is comparable to Cardiff and Miller's site responsive work in that it offers a constructed narrative about site. Site, in other words, becomes the screen for projections. The virtual imagery blurs the distinction between invented and real history, actors and real action, projection and the conditions of site. Dean and Moren hybridize the material with the virtual through the projection of images onto the materiality of site. This approach to outdoor media art contrasts with the site-specific work discussed in Chapter 4.

This examination of alternative artistic practices that make use of outdoor projection highlights the variety of site responsive work possible with the medium. The technologies do not predispose art making towards particular aesthetic practices. There are, however, "real" stakes in art making, particularly in work that strives to open an encounter with the materiality of site. The next chapter argues that while outdoor media projection may not incline art making toward particular forms of site specificity, the expansion of the virtual into the material is compatible with existing social and capital relations. Thus, site-specific artwork that strives for an encounter with the material is a resistant engagement with the medium that attempts to mount critical opposition to these relations.

1
Janet Cardiff and George Bures Miller, *Paradise Institute*, 2001

2
Janet Cardiff and George Bures Miller, *Paradise Institute* [alternate view], 2001

3
Janet Cardiff and George Bures Miller, *Her Long Black Hair*, 2004

4
Graffiti Research Lab's Mobile Broadcast Unit

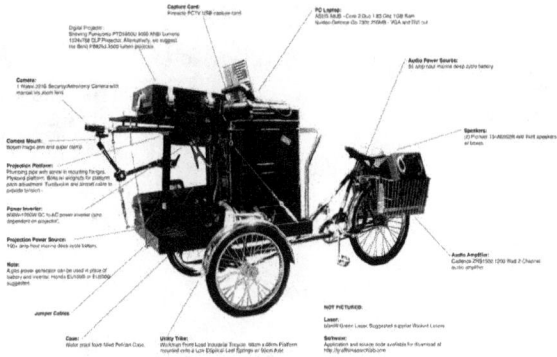

5
Minneapolis Art on Wheels' iteration of the Mobile Broadcast Unit

6
A screenshot of an early version of the custom-written software package, Livedraw.

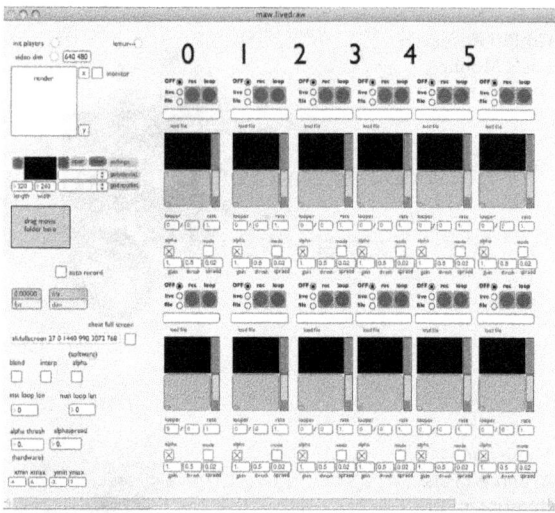

7
Minneapolis Art on Wheels, *Seaworthy*, 2010

8
Mobile Experiential Cinema, *Second Bridge is Wider, but not Wide Enough*, 2012.
Image: Ben Moren

9
Mobile Experiential Cinema planning diagram, *The Parade*, 2013.
Image: Trevor Burks

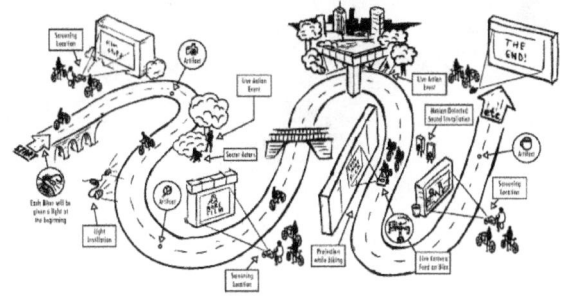

Rupture of the Virtual

2004. November 8. 2:25 P.M.

At first the road looks like it has been passed through a mesh screen. Everything is two dimensional—the sky is painted on the same plane as the trees, the buildings, the road. Minor adjustment to the eye's perception of depth allows one to pass through the screen and restore perspective. It is no different than looking through a picture window or coming out of a daydream in the passenger seat of a car.

A voice comes in over the headphones. Take a right into the trees. Nothing more until well within the boundary of the wooded area when the voice returns. The tone is now militant. Modern entertainment technologies have transformed life into a passive experience of the everyday. Today's cineplexes are designed with the efficiency of slaughterhouses. They are thinly disguised mechanisms for creating heightened levels of docility during consumption. The realization of this logic is the installation of video screens as distractions from the emergency of experience. Mounted into the seats of cars and airplanes, a passenger could care less if the sensation of descent is a pleasurable descent into Heathrow International or a crash-landing into the Atlantic. Keep walking.

A feeling that begins in the gut transforms into the sound of a distant rumbling not unlike thunder. Keep walking. The sound is continuous and starts to advance. It begins in the left ear and approaches from behind. It takes on the quality of an airplane flying overhead. A look overhead through the trees doesn't reveal anything. Keep moving. The noise continues to grow louder and louder until, in a panic, the only solution is to rip off the headphones. Relief offered by the quiet is broken by a visual riot as disorienting as the white noise. Animal alarm, a sensation unfamiliar to predators. A megashape intrudes on peripheral vision. The occlusion: a gray stain with a tapered cylindrical form. It's too large to make out any detail, but it's moving slowly overhead. As it passes, the headphones, dangling by its cord, rattle with distortion. A fin shape can be made out, which now resolves into a wing. What seems to be a rudder with working ailerons comes into view? No longer taking up the entirety of the sky, it is an airplane, flying overhead, brushing the treetops yet impossibly passing through them.

Headphones are retrieved once the noise that pumped through them stops. Mid-sentence the talk continues... But the underlying problem is the same. The wrong questions are being asked. We are not interested in recovering reality from its shroud of virtuality. We need to concern ourselves with other problems; other questions need to be asked; other issues attended to. The fixation of attention—the sedentarization of daily work and leisure life— has been realized in the diffusion of the personal computer on which both work and entertainment are performed. All time has been monopolized by the forces of production that has led to the construction of desires that connives us to stay in our seats. Docility and spectatorship are hand-in-glove in the production of today's spectator, who sits immobilized in hypnotic attention at the cinema, in front of the television, and now before the computer terminal.

A mistake has often been made in the thought that if it were possible to break the spectator's gaze from its entrancement, it would be possible to liberate reified experience. This is to say that certain types of mobility would be an antidote to docility, freedom from fixatedness. Isn't this the logic of the dérive—the roaming drift that eludes the psychological order of modern life by exploring movement outside of the control of centralized authority and dominant economic force? Mobility has been the experience promised by entertainment technologies since their early form as panoramas. It will become the ideology by which new technologies built upon physical mobility, such as the one you are using now, will be popularized. Transporting a viewer to distant locations from the pleasure of your own home has already been realized. So what if you're now walking in the woods. It's just as easy for someone who is in nature to dismiss a preternatural noise as only realistic sound effects, as it is for someone who is terrified by the sudden approach of an animal on the screen to proclaim, "It's only a movie!" The terrifying truth is that we would actually prefer that a foreign noise in the woods was a sound effect piped in for our entertainment...

Above is an excerpt of a guided walk-through of my interactive installation *Airplane Tree Minaret* (2003), which uses Augmented Reality (AR), a type of See-Through Graphical

Interface, as its medium. In an examination of this and another site-specific work, this chapter elaborates on claims already introduced in this book, specifically that site-specific media art can open an encounter with the material. This chapter focuses on the utilization of STGIs in site-specific works and examines the technical details and aesthetic possibilities of the creative application of the medium in rupturing the virtual. An ongoing discussion of mobility is central here, and I consider how mobile interfaces, such as Augmented Reality, can facilitate critical awareness of the virtual.

Airplane Tree Minaret explores AR's ability to overlay virtual objects onto site. Participants are given Head Mounted Displays with integrated headphones before entering the work, which begins when a participant arrives at the space of the installation. Audio is fed into the Head Mounted Display's headphones, a noise which crescendos in pitch and intensity. The audio—a dull rumbling similar to the sound of an approaching jet engine—is spatialized to sound like it is coming from overhead, but there is nothing in the sky that elicits the noise. When the sound reaches its peak level, as suddenly as it started, it drops to a minimum. Simultaneously, a realistic computer generated airplane flies overhead, passing improbably through the treetops (see Figure 1).

The participant interacts with the first part of the installation until her physical movements through the space trigger the second, which builds upon the sensory disorientation already created by the projections and sounds. When a participant is within 50 yards of a predetermined geographical spot within the installation, referred to as the "clearing," the second part begins. The audio heard throughout the first part of the project restarts, but this time it does not grow in intensity according to its own internal timing, but is linked to the viewer's physical proximity to the center of the clearing. The deeper the participant moves into the clearing, the louder the sound grows; farther away, the sound diminishes. It is as if there were a point in the clearing eliciting a noise that grows louder as the participant draws near, but the audio is actually generated by the wearable computer responding to the viewer's GPS location, not by any source external to the participant's position. With the participant's approach to the center of the clearing, the audio grows in intensity until it again reaches a level causing auditory distress. Once within visual range of the center, the participant sees the "minaret," a computer generated projection of a metal structure, about 30 feet tall, bearing speakers and reminiscent of a watchtower.

Before moving into an analytical discussion of *Airplane Tree Minaret*, I would like to review a second installation of mine in order to remark on what both share in terms of site and site-specificity. Created before *Airplane Tree Minaret*, *Military Bay* (2003) is a visualization of the entire United States military naval capacity set in San Francisco Bay. From publicly available data, I collected an inventory of U.S. military watercraft and gathered measurements of each type of ship. I projected full scale, three dimensional computer models of these ships onto the part of the San Francisco Bay contained by the Golden Gate, Bay and Richmond Bridges—an area of approximately 200 sq. miles (see Figure 2). The military ships not only covered much of this part of the Bay, but extended out into the Pacific Ocean past the Golden Gate Bridge. On bridges, piers and coastlines, viewing locations were indicated by sandwich boards with numbers printed on them. These provided panoramic views of the installation, though participants were free to move about the Bay for alternate vantage points.

Originally conceived in early 2003, during the United States' preparations for its most recent intervention into the Middle East, *Military Bay* is rooted in a critique of American military force and the impending invasion of Iraq. It reveals America's transition into a more aggressive, interventionist phase at the turn of the twenty-first century. By populating the Bay with weapons of destruction normally secreted away from sight, viewers were confronted with a visual display of the military's monetary and material expenditure and shocked by its scale.

The project also referred to AR's role in the virtualization of war violence by retooling a technology that was becoming an entertainment medium. Similar to the first Gulf War, the military, with spectacularized fanfare, released images from the Iraq War and broadcast the war as evening entertainment for the American public. Since the first Gulf War, military-entertainment technologies had evolved; footage from tele-operated drones, known as Predators, was simulcast to U.S.-based control stations and the homes of

American viewers. The existing popularity of computer-based flight simulators whets the appetites for such images and creates desire for remotely operated entertainment. A step beyond the grainy-green video produced by smart bombs, the high-resolution footage from these remote-controlled killing machines demonstrates the military's uses for AR. The technologies underlying tele-operated drones and AR are comparable: in both, a human actor operates in a geographical environment where information and action are mediated through a computer display.

Military Bay was more than its political critique of military power and its assessment of AR as a military-entertainment technology. On a different level, the installation unsettled participants by taking the site (the Bay), which had fixed associations for local residents, and turning it into something foreign or unfamiliar. The project colonized a space normally protected from military displays and used for water sports, tourism and commerce. The projection of virtual objects onto the Bay transformed one's relationship to the site. What changed, however, was not the site itself, but one's perceptions of and associations with it.

MATERIAL SPACE INTRUDES: THE RUPTURE OF THE VIRTUAL

Airplane Tree Minaret and *Military Bay* both interrupt a viewer's experience of place, but do so in ways that make obvious the artifice of their projections. Computer generated military ships mysteriously float on the surface of the water unresponsive to the motion of the waves. Meanwhile, waves impossibly break through virtual objects. Airplanes fly through real treetops and boats pass through the renderings of military watercraft. These constitute "imperfections" in the projections. Hito Steyerl is fascinated with the imperfect or "poor" image, because it reveals itself as a "digital uncertainty."[161] These imperfections must be understood not simply as the image's inability to live up to the quality of the original, rather the revealing of digital processes that have conditioned the image in this way (through compression, data errors in digital copying, etc.). In *Military Bay* and *Airplane Tree Minaret*, a digital uncertainty exists at the difference between the virtual and the material. This uncertainty is not a limitation, but an opportunity for site to interrupt projection.

As discussed in Chapter 4, STGI-based artwork has the potential to rupture the virtual by opening an encounter with the material. This was articulated in Krzysztof Wodiczko's projection of images onto the built environment that foregrounded site's materiality. A recognition of the material from behind the screen of the virtual distances viewers from an embeddedness in the virtual and invites an interrogative address of the virtual. *Military Bay* and *Airplane Tree Minaret* both rupture the virtual by producing a digital uncertainty that enables the material to crash through the virtual.

A second way in which these projects rupture the virtual is the manner in which they employ mobility. Viewers can circumambulate Wodiczko's projections, but his work is stationary. Viewers have restricted viewing locations from which to view Wodiczko's work. *Military Bay* and *Airplane Tree Minaret*, by contrast, require physical mobility in order to view the pieces. The viewer must move through and around the site to experience them. As I argue next, mobility and its simulation have been central to the development of modern media, and by looking at the role that virtual mobility has had in the disciplinary production of the modern docile subject, we can recognize how a digital economy has enforced attention and sedentariness. Site-specific installations that encourage physical movement can open an encounter with the material through a disruption of this enforced attention and sedentariness.

THE SPATIALIZATION OF THE INTERNET AS A VIRTUAL FIX

The Internet abounds in spatial metaphors that promise to take us on "virtual trips," "fly-throughs," "global road trips," etc. All this without leaving the comfort of our chair. For example, Hyperlapse, a website released in 2013, automates the process of stitching together Google's Street View photographs to simulate cinematic fly-throughs between any two locations. These fly-throughs create a strangely disembodied experience where you float slightly above eye-level[162] skimming the roofs of cars and heads of pedestrians. It is the pure visual experience of speed; tourism is no longer about arriving at a destination, but the pure sensation of flying over roads and highways.[163] Though Street View photographs are of real locations, one's movement is in no way a physical experience, but one fabricated by the hardware and software infrastructure that comprise the Internet and computer graphics. One selects the destinations one wants to travel between on Google Maps (locations which exist as representations of places on a computerized two-dimensional map) and travels along a route chosen and constructed by the Hyperlapse algorithm. While roller coasters produce a sensation of speed by catapulting our bodies through the air around, Hyperlapse accomplishes this through the visualization of diverse software and hardware processes as spatial form.

Hyperlapse is an example of what Wendy Chun and others have characterized as the spatialization of the Internet, that is, the representation of the Internet with spatial characteristics. This spatialization is remarkable because it has no basis in the design or structure of the Internet. In an analysis of the communication protocols and hardware that structure Internet communications (such as TCP/IP), Chun underscores the differences between the Internet's construction and actual physical space. The dangers in presuming that the Internet has a spatial dimension belies how fundamentally dissimilar they are. While we have a relative freedom to roam in physical space, the layers of protocol that comprise Internet communication make information subject to microforms of control and surveillance.[164]

The crucial question, however, is not why the spatialization of the Internet is preserved, despite a recognition that there is "no space in cyberspace,"[165] but what ideological function spatialization serves. I argue that this spatialization has contributed to the creation of the Internet as, what David Harvey has referred to, a spatio-temporal fix, that is, a labor and capital investment to redress the chronic problem of overaccumulation in capitalism.[166] While Harvey has focused on spatio-temporal fix's geographical consequences, the expansion of capitalism through the Internet can be understood as its virtual counterpart. After a discussion of the expansion of the Internet as a type of spatio-temporal fix, I return to a discussion of its ideological function in the production of docile bodies and the role that mobility can play as a critical response to this problem.

The spatialization of the Internet has been central to the rhetoric surrounding the Internet since its early popularization. In "The Wild, Wild Web: The Mythic American West and the Electronic Frontier," Helen McLure identifies the spatialization of the Internet in a comparison of the characterization of the Internet as an e-frontier to the rhetoric that helped to inspire the colonization of the American West. She notes that both were conceived as open spaces, ripe for expansion and investment:

Elements of space and time define and link both the American western frontier and the global e-frontier. [Frederick] Turner's "moving frontier" encompassed both physical space and time and was distinguished by its apparent closure by 1890; in contrast, the e-frontier is also spatial and temporal but is usually imagined as new and still wide open territory.[167]

During the American westward expansion, prosperity and health were characterized in spatial and temporal terms. (Horace Greeley's famous urging: "Go West, young man, go West.") By moving West, settlers could go forward into the future and find happiness and prosperity unattainable to them in the corrupted East. A similar frontier mentality is present in the hype surrounding the Internet in works including Howard Rheingold's *The Virtual Community: Homesteading on the Electronic Frontier*, Bruce Sterling's *The Hacker Crackdown: Law and Disorder on the Electronic Frontier*, Katie Hafner and John Markoff's, *Cyberpunk, Outlaws and Hackers on the Computer Frontier*, and John Perry Barlow's "Declaration of Independence of Cyberspace."

McLure's geographical and spatial analysis, however, stops short of offering an examination of why a spatial metaphor governed the rhetoric around the Internet, despite the fact that there is no space in cyberspace. David Harvey's work on the spatio-temporal fix offers us a way to conceptualize it. Following Marx, Harvey observes that capitalism is prone to periodic crises due to inherent contradictions in the capitalist economy. There are a number these, but the contradictions that relate directly to the spatio-temporal fix concern the role of surplus of capital and labor power. Capitalists continually seek to increase their access to capital through a mix of tactics, such as technological innovation, education, the expansion of access to a cheap labor force, etc. An outcome of this drive for expansion is chronic surpluses of capital and labor. Harvey refers to this situation as overaccumulation, or "the condition in which too much capital is produced relative to the opportunities to find profitable employment for that capital."[168]

Harvey has identified two historical responses to the problem of overaccumulation: temporal deferral or displacement and the spatial fix (together, the spatio-temporal fix). Harvey writes:

A certain portion of the total capital is literally fixed in and on the land in some physical form for a relatively long period of time (depending on its economic and physical lifetime). Some social expenditures (such as public education or a healthcare system) also become territorialized and rendered geographically immobile through state commitments. The spatio-temporal "fix", on the other hand, is a metaphor for a particular kind of solution to capitalist crises through temporal deferral and geographical expansion.[169]

Temporal deferral consists of large-scale investments in infrastructure and projects whose value take long periods to realize, if ever. (This can include roads and highways, large-scale energy projects, social infrastructures of education, healthcare, social services, as well as investments in state administration, law enforcement and military protection.[170]) Temporal deferral displaces the problem of excess surplus capital to the future by dumping it into infrastructure. The spatial fix, on the other hand, operates in conjunction with temporal displacement, for temporal displacement can occur by "scale enlargement," that is, by expanding into new regions and spaces for capital expenditures.

Harvey's spatio-temporal fix is built upon Marx's insight that "the need of a constantly expanding market for its products chases the bourgeoisie over the whole surface of the globe. It must nestle everywhere, settle everywhere, establish connections everywhere."[171] It is a mechanism for the creation of the necessary infrastructure for regions to be integrated into a global capitalist economy. China's integration into global capitalism is the most notable spatio-temporal fix of the last few decades. With investments in infrastructure (roads, dams, universities, etc.) in the hundreds of billions of U.S. dollars, China has "the potential to absorb surpluses of capital for several years to come."[172] With this investment in infrastructure, the spatio-temporal fix conjures into being new cities while redeveloping old ones as well. This is why Harvey refers to an outcome of the spatio-temporal fix as the capitalist "production of space."[173]

A spatio-temporal fix is also in evidence in governmental and corporate financing of Internet hardware and software technologies, which has been well-documented. The U.S. government provided funding to the Advanced Research Projects Agency (ARPA, later DARPA) for the invention of the communication protocols integral to the Internet. Estimates put the cost on the low end of around $25 million from 1969 - 1990.[174] Vint Cerf, ARPA's program manager on Internet related projects during that period, puts the figure closer to $50 - $70 million.[175] As should be expected given Harvey's claim that the spatio-temporal fix is a response to financial crisis, the U.S. government continues to fund the Internet directly and indirectly at a level much higher in times of crisis and recession. In 2009, as part of a major stimulus package intended to redress a global economic decline, the federal government allocated $7 billion to high-speed Internet communications for the installation of broadband access for educational programs and underserved neighborhoods.[176]

These amounts are meager in comparison to private capital investments in Internet infrastructure, hardware and software, including fiber optic lines, cellular phone service networks, data centers, software, etc. There is no reliable estimate for the total amount of funding put into the Internet by private companies but, given the number and wealth of businesses involved in provisioning Internet service, this number would be enormous, far eclipsing government funding. Google's capital investments on servers, data centers and other infrastructural improvements alone exceed $1 billion per quarter[177] and Google is just one company among many that drives money into Internet technologies.

From an economic perspective, the spatialization of the Internet encourages its utilization as a spatio-temporal fix for surplus capital and labor. Capital expenditures in and around the Internet have been long-term investments in the reinvention of capitalism. The Internet has become a digital platform for the unprecedented mobility and a playground and factory for the harnessing of novel forms of labor power and the creation of capital.[178]

DOCILE BODIES
AND VIRTUAL
MOBILITY

Even before the advent of the Internet, the media had been rendered in spatial terms. Oliver Grau and Anne Friedberg[179] both read the history of modern media as the construction of virtual spaces of illusion. Friedberg traces the origin of such virtual spaces to an unlikely location, the nineteenth-century French shopping arcade, which, for Friedberg, was a precursor to modern media technologies. She compares the cinema-goer to the figure of the flâneur, whose aesthetic pleasure moving through halls of the arcade, is akin to a cinematic-type experience. As the flâneur wanders through the arcade's displays and spectacles, so too do media technologies (such as the panoramas, cinéoramas and, eventually, the cinema) promise viewers transport or mobility through virtual spaces.[180]

What distinguishes modern media technologies from the spectacle of the arcade, however, is that actual physical movement is no longer constitutive of the experience of mobility. Whereas the flâneur ambled through the halls of the arcade, today we sit in fixed attention to the screen and experience virtual spaces in the comfort of our chairs. Though mobility is integral as an aesthetic dimension to modern media technologies, such as cinema, it has become, concludes Friedberg, a "virtual mobility," insofar as modern media entrain the viewer to the screen and do not necessitate actual physical movement.

Jonathan Crary's study of the disciplinary mechanisms that have led to the construction of the modern media spectator complements Friedberg's account of virtual mobility.[181] In fact, Friedberg's notion of virtual mobility would be strikingly incomplete without Crary's insistence that an individual's capacity to stare in fixed attention to a screen has had a contingent relationship to the development of apparati of discipline and control in the construction of the modern spectator. This is to say that the experience of virtual mobility simultaneously frees the viewer's imagination from the here and now, but paradoxically chains him to his seat. This insight is precisely what is at stake in Crary's research on the history of attention. He writes:

Many of the modes of fixation, of sedentarization, of enforced attentiveness implicit in the diffusion of the personal computer have achieved some of its disciplinary goals, in the production of what Foucault calls docile bodies. The proliferation of electronic and communication products ensures that docility will always be linked to intensified patterns of consumption.[182]

Extending Crary's observations about docile bodies and patterns of consumption to the research on digital labor, we can argue that virtual mobility is not only integral to patterns of consumption (of media such as video games, films, television, etc.), but to work in the digital economy as well. During the industrial era, Frederick Taylor scrutinized the movements of assembly line workers to extract greater productivity.[183] Today, management analyzes the attentiveness and movements of information workers with precision in order to extract greater efficiencies in screen time, typing speed, and time-at-keyboard. Crary's account of disciplinary attention is hand-in-hand with Friedberg's account of mobility: the Internet's spatialization contributes to the production of sedentary workers in a screen-based economy. Any discomfiture with the prolonged docility of modern life is sublimated in the dream of virtual mobility, an ideological fantasy of freedom from fixation and enforced sedentariness.

Repression and sublimation, however, are not the only responses to media culture. Art movements have long embraced movement and mobility as forms of critical resistance to the disciplinary chains of a media-based economy. The next section reviews the preoccupation with physical movement in avant-garde art as a critical artistic practice for resisting docility and spectatorship. This will enable us to return to a consideration of the role of mobility in site-specific media art as the critical experience of space.

PHYSICAL MOVEMENT AS A STRATEGY OF DÉTOURNEMENT

The Situationists were a European radical artistic and intellectual collective that existed in various forms from 1957 to 1972. Producing a body of critical writings and artistic interventions, they sought to provide a "radical critique of modernist art practice, a politics of everyday life, and an analysis of contemporary capitalism."[184] The dérive, often translated as the drift, was a key Situationist aesthetic tactic to upend mid-twentieth century European bourgeois society. It consisted of a wandering through a city on foot, "a drifting without destination, neither going to work nor properly consuming, a waste of time in the temporary economy."[185] Lasting anywhere between one hour to months at a time, while on a dérive, the drifter could happen upon obscure locations, places of ambiance, forgotten landmarks, and encounter unfamiliar people and places.

There was a critical method informing the dérive in spite of its apparent aimlessness. In his "Theory of the Dérive," Guy Debord confidently defined it as "playful constructive behavior," which "changes our way of seeing the street."[186] By compelling participants to break out of habitual movements through a city, the dérive "alerts people to their imprisonment by routine."[187] Evidence that we have become sites of discipline and control, according to Debord and the Situationists, is in our bodily navigation through a city. The route we take from point A to point B has been chosen for us by a multitude of disciplinary forces, such as urban planners, who have designed routes according to the greatest amount of surveillance and control; by capitalist pressures for the maximization of efficiency; by our attraction to spectacles, such as the pretty landmark by which we like to pass as a morning ritual. What an individual on a dérive becomes aware of in his rootless wanderings is how behaviors are unknowingly governed by the convergence of these disciplinary mechanisms. Social and capital relations are inscribed and articulated in our very movements through a city, and they get brought to the surface through participation in the dérive. More than just a tactic for recognizing the existence of disciplinary mechanisms, the dérive was a revolutionary practice for breaking free of such forces of control.

Military Bay and *Airplane-Tree-Minaret* also employ mobility as a basis for their effect. Whereas the dérive encouraged participants to take over the city as a way of overturning disciplinary mechanisms, these installations utilize mobility in order to orient the viewer to structures that have come to dominate our everyday interactions. In contrast to the disciplinary goals of media technologies to create docile bodies, STGI installations encourage viewers to wander the space of the installation and linger in contemplation of the relation between the project and site, the site and their physical location. When viewers of the *Military Bay* installation, for example, travel to locations around the San Francisco Bay—frequented predominantly by tourists seeking panoramic views of the city—they do so not because of the attractions of spectacle, but the re-envisioning of the space as a site documenting military power. The space of the installation is appropriated by the project and re-configured in the minds of viewers: the Bay is no longer the site of leisure activities, commerce or spectacle, but the space of possibility, loosened from the disciplinary mechanisms that construct the city for its inhabitants. Is this not a realization of Debord's rationale for the dérive as a strategy of détournement? For participants to "[drop] their usual motives for movement and action"?[188]

No longer fixed in rapt attention to the screen, people move through and around their surroundings to encounter space as a site of possibility and meaning. The installations employ physical mobility as a way to enable viewers to envision material space as centers of "possibilities and meaning" not dominated by ingrained psychological patterns and fixed social and capital relations, which cement the experience of a particular locality.[189] In this way *Military Bay* and *Airplane-Tree-Minaret* loosen the disciplinary goals of modern spectatorship and open alternate ways of experiencing one's surroundings.

In sum, site-specific STGI-based media

art opens an encounter with site and its materiality in multiple ways. It draws attention to the materiality of site through the formal differentiation of the virtual (as media image) and the material (as surface of projections). This differentiation is a digital uncertainty when the material crashes through the imperfection of the projections. By extension, this uncertainty produces an awareness of the material as the screen for projections. A second way that STGI-based media art opens an encounter with the materiality of site is by challenging the disciplinary forces that proscribe one's interactions with site. When employed as a critical artistic practice, mobility is an effective response to the disciplinary forces required by a digital economy. Against the reproduction, normalization and expansion of forms of fixed attention and sedentariness in labor and leisure time, media art literally compels viewers to get out of their seats and attend to the presence of site beyond the screen.

In its capacity to open an encounter with the material, site-specific media art enables a critical recognition of the virtual as a cultural condition. Because STGIs construct experiences situated at the intersection of the material and the virtual, it is a medium that is well suited to the formation of such a position. Our interactions with site are conditioned by virtual processes, and this habitus requires disruption before such a position can be won. Site-specific media art can provide such a disruption, because it can enable a recognition of the presence of the material. An awareness of the material in the experience of site is necessary for the articulation of a critical position regarding the virtual. Together, this recognition and critique constitute a rupture of the virtual. The next chapter, the Conclusion to this book, extends these claims by exploring how varied cultural practices outside the realm of art can also produce ruptures of the virtual.

1
John Kim, Site-specific Augmented Reality Installation, *Airplane Tree Minaret*, 2003

2
John Kim, Site-specific Augmented Reality Installation, *Military Bay*, 2003

Conclusion

In the last decade there has been a marked proliferation of art and cultural events that are organized around media projection and outdoor screening. In outdoor film events, digital billboards, guerilla video bombing, interactive projection, video walls, etc., projections have moved from darkened theaters to public squares and parks, backyards, the sides of buildings, and a myriad of other outdoor spaces. *Nuit blanche* art festivals, such as Minneapolis's *Northern Spark*, Toronto and Paris's *Nuit Blanche*, Santa Monica's *Glow* and others, have become major venues for outdoor projections and public media installation art. This expansion has spawned research initiatives, frequently referred to as "urban screens," and the growth of outdoor projections is heralded by advocates and researchers for their capacity to build "community, local identity and engagement."[190]

Against the perceived benefits advocated by their supporters, this book has suggested that it is difficult to speak of a singularly positive function for outdoor projections, especially one that could broadly result in the creation of community, localism or, even, identity. Indeed, a common use of site in outdoor projection is simply to add ambiance or mood to events, and in this sense, rather than enhancing localism or community, the media can further distance us from our built environment by projecting the virtual out into it. Site recedes with the dimming of lights, regardless if the screen is indoors or out.

Work that strives to reference and interrogate the material, like the examples of site-specific media art reviewed in this book, pose an alternative model for media projection, one that takes seriously a desire to engage viewers in and *about* their surroundings. In the face of proliferating ways to extend and enhance virtual experience, we must attend to practices that draw attention to the material, especially as a critical address of the virtual in our lives. Neither retrograde nor nostalgic, such media practices can provide us with an awareness of our cultural condition and connect us to the social meaning of site and its materiality.

In this book I have developed a conceptual and critical framework with which to recognize and interpret media practices that strive to open such connections. The material's exclusion from media research is reinforced by the proliferation of types and sizes screens, which offer up more immersive virtual spaces. The material's exclusion has been fundamental to research on the virtualizing forces of contemporary culture. This book is obviously indebted to this area of research, but I protest against the focus on technologies of the virtual as prosthetic extensions of vision. I have argued that the media have also had a historical and ongoing role in constructing a relationship with those things that are in front of our noses, that is, our immediate physical surroundings. The utility of this concept has been documented through the study of the historical development of media and computing perceptual devices that help cognize our immediate surroundings.

This book has mainly drawn on examples from media art in making its claims about the material. Beyond these particular examples, it is possible to recognize a preoccupation with the material in cultural practices outside of the realm of art. In order to make such a claim, I want to first recall arguments made earlier regarding the material. In Chapter 5, I noted Janet Cardiff and George Bures Miller's appropriation of the Walkman to encourage "a sense of connection to place." This need for connection to place can be read as symptomatic of a range of problems in contemporary life that contribute to a sense of separation from an intimacy with our surroundings. Dan Schiller, for example, has written about the Walkman as an example of our growing separation and disconnection from others and our environment, what he refers to, drawing on Raymond Williams, as "mobile privatization."[191] Personal media technologies contribute to a technologically mediated sensory bubble that shields us from contemporary life by keeping it at an imagined arm's length. From personal entertainment devices, like the Walkman, iPod and smartphones, to the rebooting interest in Virtual Reality, we live in the midst of people, places, and histories, but feel a sense of disconnection from them.

Manuel Castells,[192] Sherry Turkle,[193] and others have maintained that this sense of disconnection and dislocation is constitutive of life today, because lived experience is everywhere striated by capital and media flows. While media technologies like the Walkman, the computer screen, digital media, and others, are influential in the production of life's dislocations, there is nothing inherent to the logic of their construction that determines their particular function, just the accretion of historical precedents that constrain the imaginative uses that we conceive for them. As Cardiff and Miller reveal in their work, we can claim alternative uses for media devices: instead of privatization and separation, the Walkman can be employed for empathy, memory and situatedness. Cardiff and Miller's use of the Walkman and comparable technologies is a resistant engagement with the medium; they strive to employ such devices in the reverse of how they are frequently employed.

Countervailing pressures are similarly poised against the class of technologies I have referred to as See-Through Graphical Interfaces. As discussed in earlier chapters, there is commercial and military pressure to utilize the interface for particular ends, such as the exertion of knowledge and control over the material and the expansion of the virtual into our experience of the material. Google Glass is the most visible example of this tendency in recent years. As suggested above, however, resistant engagements with the media are possible despite the pressures of capital and mass entertainment. Site-specific media artists have articulated practices that employ See-Through Graphical type technologies for purposes contrary to how they are being managed by businesses. My own work with the medium has been inspired by the envisioning of alternative possibilities for STGIs in opposition to growing normative pressures that constrain their imaginative possibilities.

CULTURAL PRACTICES AS ENCOUNTERS WITH THE MATERIAL

The latter half of this book's investigation of media practices that open encounters with the material focused on art to clarify a set of tendencies concerning the material and efforts to bring it to critical consciousness. I focused on three distinct ways in which media art have a role in producing knowledge of the material. First, in a recognition of the sociocultural conditions of site, or an understanding of the social or cultural conditions of a particular location; second, in an awareness of the disciplinary function of site, or site's disciplinary role in the formation of individual subjecthood; and finally, as an encounter with the materiality of site through a recognition of the formal distinction between the virtual and the material. I suggested that such cultural practices are realized in a critical recognition of the virtual as a cultural condition. I referred to such practices as enabling a rupture of the virtual.

This book's almost exclusive focus on art does not presume to suggest that it is the only cultural practice that opens encounters with the material. It is possible to identify a wide range of practices that do this as well. To briefly consider one example here, I am interested in how the growing urban farming movement (pursued both as a collective and an individual activity) is an instance of a cultural practice outside of the domain of art that can also be understood as opening an encounter with the material. Though it is impossible to generalize across the diverse motivations in peoples' desire to cultivate, a sentiment one can find in the movement is that growing one's own food is a form of protest against the lack of knowledge about the conditions in which industrial food products are grown and made. (Among many issues of concern, this includes the use of pesticides, herbicides and fertilizers in industrial farming practices.) The experience of packaged foods in supermarkets is itself an idealized representation of the food items themselves. They are virtualized products; their packaging masks their organic origin. Industrial food products are promoted, furthermore, in ways that embed them into spectacularized channels of advertising and consumer capitalism, separating their food content from the conditions in which they are grown.

The argument about the material offered in this book offers a distinctive way to conceive of urban farming. The movement can be understood as a reaction against the forms of mediation in contemporary life that extend down to the purchasing and consumption of food products. Taking control over the means of production re-materializes food in such a way that the social and environmental conditions in which they are grown can be revealed. (In many urban plots, one must test the soil. The discovery of harmful contaminants, such as lead or arsenic, uncover the industrial history of the site. Testing reveals the social history of the land.) Farming is also a material practice in its tactility. Hands are rubbed raw from pulling weeds and using farm tools; fingernails get caked with dirt, and fingertips feel the dampness of the ground. The formal qualities of the materiality of the earth are discovered in the practice of farming. This is a materiality that contrasts to the virtualized experience of shopping in supermarkets. A wider net needs to be cast to discover cultural practices that open encounters with the material in ways that differ from art's aesthetico-critical tactics.

Finally, this book traced the origin of the See-Through Graphical Interface to World War II military technologies. Cinema did not simply have its origins in the turn of the nineteenth century with the invention of devices for playback, recording and film processing, but in longer historical transformations that established changes in visual organization and perception. It is these changes that cleared the way for film's development. Similarly, this book has implications for the history of See-Through Graphical type technologies that originates even before the first half of the twentieth century. In its re-presentation of a viewer's physical surroundings, the camera obscura, for example, has frequently been associated with the rise of technologies of the virtual. The device, however, clearly connects the viewer relationship to the material as well. In other words, what was the camera obscura's role in situating the viewer in relationship to her immediate physical surroundings? A comparable question can be asked of early portraiture and landscape paintings. What was the function of landscape paintings that would hang for decades in the environment they depicted? Questions, such as these, remain to be answered in future research on the subject.

Notes

1. Caroline Jones, *Sensorium* (2006), 5.
2. Marshall McLuhan, *Understanding Media* (1964).
3. See Rob Shields's (2003) expansive definitions for the virtual.
4. Anne Friedberg, *Virtual Window* (2006), 11.
5. Anne Friedberg (2006) refers to the Graphical User Interface as a virtual window onto spaces of illusion that can have little corollary to spaces that exist in the physical world.
6. The lexical opposition is really between mediated and its negation (immediate), but as is discussed next, the immediate and medium are frequently conceived of as opposing twins. Given the media's relevance to this distinction, it is appropriate to consider this one, which I argue next is also conceived of in opposition.
7. Raymond Williams, *Marxism and Literature* (1992), 158.
8. Ibid, 159.
9. See "immediate" in the Oxford English Dictionary (2014).
10. Diderot, "Salon of 1763" in Eitner, *Neoclassicism and Romanticism (1750-1850)*, 58.
11. Jacques Derrida, *Dissemination* (1976).
12. Ibid, 127.
13. Jay Bolter and Richard Grusin, *Remediation* (1999).
14. Ibid, 47.
15. See Jean Baudrillard (2008) and Mark Hansen (2006).
16. Paul Virilio, *War and Cinema* (1989), 88.
17. Jason Farman, *Mobile Interface Theory* (2012). Gabriella Giannachi and Nick Kaye, *Performing Presence* (2011).
18. One example of such a use is the company Occipital (http://occipital.com/) and the development of AR technology for tablets.
19. Stanley Cavell, *The World Viewed* (1971), 24.
20. Friedberg, *Virtual Window*, 11.
21. See Jason Farman, *Mobile Interface Theory: Embodied Space and Locative Media* (2012); Steve Benford and Gabriella Giannachi, *Performing Mixed Reality* (2011); and Gabriella Giannachi and Nick Kaye, *Performing Presence: Between the Live and the Simulated* (2011).
22. Jacques Derrida, *Of Grammatology* (1976).
23. As Shields, Lister, and others have suggested, there are a multitude of definitions for the virtual across different areas of research. The intention here is not to provide a review of all of them.
24. Friedberg, *Virtual Window*, 7.
25. As is reviewed later, this did not deter Friedberg from submitting her own definition for the term as the "liminally immaterial."
26. Jacques Derrida, *Specters of Marx* (1994).
27. Ibid, xviii.
28. Ibid, 13.
29. Ibid, 54.
30. Mark Poster, *What's the Matter with the Internet* (2001).
31. Jean Baudrillard, *Perfect Crime* (2008), 35.
32. See Anne Friedberg, *Window Shopping: Cinema and the Postmodern* (1994); and Oliver Grau, *Virtual Art: From Illusion to Immersion* (2004).
33. Jay Bolter, et al., "New Media and the Permanent Crisis of Aura" (2006), 22.
34. Friedberg (2006) does consider the influence of the virtual over our physical environment in her examination of how architectural details came to mimic forms found in cinema, such as the use of horizontal windows in buildings that referenced cinema's wide format projections. But her analysis is less an account of the material, than an example of

the virtual's influence over ideas and forms found in our physical environment.
35 See Chapter 4 for a more detailed discussion of Farman's analysis of Augmented Reality.
36 Farman, *Mobile Interface Theory*, 43-44.
37 Mark Hansen, *Bodies in Code* (2006), 142.
38 Katherine Hayles, *Electronic Literature* (2010), 48.
39 Katherine Hayles, *How We Became Posthuman* (1999), 193.
40 Ibid, 208.
41 Anna Munster, *Materializing New Media* (2006).
42 Hansen, *Bodies in Code*, 5-6.
43 Donna Haraway, *Simians, Cyborgs, and Women* (2001), 149.
44 Hansen, *Bodies in Code*, 59.
45 Ibid, 123.
46 Ibid, 5.
47 Judith Butler, Ernesto Laclau and Slavoj Žižek, *Contingency, Hegemony, Universality* (2000), 108.
48 Judith Butler, *Gender Trouble* (2006).
49 Sarah Whatmore, "Materialist Returns" (2006), 600-609.
50 Arthur Schopenhauer, *The World as Will and Representation* (2010), 50.
51 Paul Milgram and Fumio Kishino, "Taxonomy of Mixed Reality Visual Displays" (1994), 1321.
52 Ronald Azuma, "A Survey of Augmented Reality" (1997).
53 Steve Mann, "Wearable Computing as Means for Personal Empowerment" (1998).
54 Nathan Jurgenson, "Digital Dualism versus Augmented Reality" (2011).
55 Ronald Azuma, et al. "Recent Advances in Augmented Reality" (2001).
56 Manovich, *The Language of New Media*.
57 Farman, *Mobile Interface Theory*, 39.
58 Farman, *Mobile Interface Theory*, 43-44.
59 Michelle Tan, "Pioneering Stryker Unit Preps for Afghanistan" (2009).
60 *Commerce Business Daily* (1993).
61 Chris Gray, *Postmodern War: The New Politics of Conflict* (1997).
62 Ibid, 196.
63 As Lenoir (2003) has observed, military research and development of new technologies is frequently a joint military-educational-industrial venture. This is true for the Land Warrior program. The various institutions that have received governmental funding through the program include Columbia University, University of Washington, Microvision, General Electric, General Dynamics, Defense Advanced Research Projects Agency, and others.
64 Though the analysis offered here focuses on the combat foot soldier, it is worth mentioning that comparable research and development programs have been aimed at upgrading visual computer interfaces across military vehicles and hardware. Head-Mounted interface devices are already found in aircraft and land vehicles, some of which I discuss below. For an account of the military's recent involvement in developing the Head-Mounted Display, see Grau (2004: 163) and the US Army's conceptual framework doctrine entitled, "AirLand Battle 2000."
65 General Dynamics (2010).
66 Virilio, *War and Cinema*, 88.
67 See Livingston et al. (2003); Defense Advanced Research Projects Agency (2008); Cameron (2010).
68 Livingston et al., "An augmented reality system for military operations in urban terrain" (2003).
69 For studies of the military's role in developing early computing technologies, see David Mindell's Between Human and Machine

70 Caudell and Mizell (1992) observe this basic similarity in their work on early Augmented Reality devices, yet they do not recognize the historical and theoretical significance of this connection.

71 Daniel Weintraub and Michael Ensing's anecdotal history of the HUD is noteworthy for its astute acknowledgement of J.M. Naish as a "prime mover in the conception and development of head-up displays" (1992: i). As I document in this chapter, Weintraub and Ensing, however, incorrectly identify Lt. Col. Paul M. Fitts as responsible for the earliest reference to the 'head-up display concept' (1992: 3) in 1946. I argue here that the Head-up Display has its origin in the design of the Gyro Gunsight in 1942.

72 JM Naish, "A study of O.R. 3044 for a navigation, bombing reconnaissance and flight control system for strike reconnaissance aircraft" (1958), 11. Quotation marks in the original.

73 Ibid, 36-37.

74 Ibid, 36.

75 JM. Naish, "The head-up steering display for the T.S.R. II aircraft" (1959).

76 In a research paper on the design of a "Head-up Steering Display," Naish refers to gunsighting technologies: 'It would then be possible to make use of a known technique of flight, viz. gun-aiming, in which both visual and instrument information, i.e., target and market, are used concurrently, but with the intention of replacing the target by the visual environment of the aircraft' (1959: 4).

77 Archibald Sinclair, "Letter from Archibald Sinclair, Air Ministry, to J.J. Llewellin, Ministry of Aircraft Production" (1942).

78 R. Wallace Clarke, " Drawing a Bead: Part 3" (1983); and *Price, Late Marque Spitfire Aces: 1942-45* (1998).

79 A comprehensive history of the Gyro Gunsight, one that accounts for its significance in the development of interface technologies for early computing devices, has yet to be written. R. Wallace Clarke's impressively expansive British Aircraft Armament is an invaluable resource for researching the historical origin of Gyro Gunsight in the context of military gunnery and sighting devices, but his account is unreliable in places. It is insufficiently documented, and some of its historical details and claims I have found to be inaccurate. I have avoided citing passages from his texts, but used a figure from his book.

80 Royal Aircraft Establishment at Farnborough, "Gyro gunsight adoption by U.S.A.: development, and firing trials in America" (1941).

81 Science News Letter, "Gunsight for Planes" (1944).

82 Trafford Leigh-Mallory, "Pilot's gyro gunsight Mark II" (1943).

83 Timothy Moy, " *War Machines: Transforming Technologies in the U.S. Military, 1920-1940*" (2001), 50.

84 Royal Aircraft Establishment at Farnborough, "Gyro gunsight adoption by U.S.A.: development, and firing trials in America" (1941).

85 R. Wallace Clarke, "Drawing a Bead: Part 3" (1983); *Price, Late Marque Spitfire Aces: 1942-45* (1998).

86 R. Wallace Clarke, "Drawing a Bead: Part 3," 202.

87 WS Farren, "Letter to the Ministry of Aircraft Production" (1941).

88 These specific comments were in reference to a non-gyro reflector gunsight, but it was held that all fighters which employed reflector gunsights could benefit from this modification (Armament Division, 1942).

89 Ibid.
90 Wing Commander (name unknown), "Letter from the Wing Commander of the Air Fighting Development Unit" (1943).
91 Instrument & Photography Department, "Bullet proof screen as gunsight reflector" (1943).
92 Krzysztof Wodiczko and Duncan McCorquodale, *Krzysztof Wodiczko* (2011), 51.
93 Ibid.
94 Ibid, 146.
95 Ibid, 148.
96 Sigmund Freud, *The Standard Edition of the Complete Psychological Works of Sigmund Freud* (1955), 247.
97 Wodiczko, *Krzysztof Wodiczko*, 149.
98 Ibid, 172.
99 Sigmund Freud, *The Standard Edition of the Complete Psychological Works of Sigmund Freud*, 226.
100 I develop this claim in greater detail later, when I quote Anne Friedberg (2006), who makes this point regarding screen based media projections. In addition, Derrida (1994) also draws a connection between the virtual and ghosts in his examination of the specters of contemporary politics and culture in his Specters of Marx. For the sake of brevity, I refer to screen based projections as immaterial, instead of liminally immaterial.
101 Friedberg, *Virtual Window*.
102 Miwon Kwon, *One Place After Another: Site-Specific Art and Locational Identity* (2004).
103 Drawing on the research of John Beardsley, Donald Thalacker, and Harriet Senie, Kwon's definition for public art traces its origin to the 1960's modern public art movement in the United States, in which art was sited "outdoors or in locations deemed to be public primarily because of their 'openness' and unrestricted physical access—parks, university campuses, civic centers, entrance areas to federal buildings, plazas off city streets, parking lots, airports" (Kwon 2004: 60). A definition of public art I share in this chapter.
104 In the next section I consider Richard Serra's *Tilted Arc*, a site-specific work of public art made famous by the controversy surrounding attempts to relocate it, despite the fact that it was a site-specific work created for the location in which it was originally installed.
105 Rosalyn Deutsche, "Architecture of the Evicted" (1990), 31.
106 Wodiczko and McCorquodale, *Krzysztof Wodiczko*, 149.
107 Ibid.
108 Deutsche, "Architecture of the Evicted."
109 Neil Smith, "Contours of a Spatialized Politics: Homeless Vehicles and the Production of Geographical Scale" (1992).
110 Wodiczko and McCorquodale, *Krzysztof Wodiczko*, 104.
111 Ibid, 104.
112 Kwon, *One Place After Another*, 60.
113 Ibid, 74.
114 Ibid, 57.
115 Ibid, 74.
116 Jeffrey Skoller, *Shadows, Specters, Shards: Making History in Avant-Garde Film* (2005).
117 Kwon, *One Place After Another*, 65-66.
118 Krzysztof Wodiczko, *Critical Vehicles: Writings, Projects, Interviews* (1999), 47.
119 Krzysztof Wodiczko, "Projections" (1990), 284.
120 Ibid, 284.
121 Louis Althusser, *For Marx* (2005).
122 Hal Foster, *Discussions in Contemporary Culture* (1987), 42.
123 Wodiczko and McCorquodale, *Krzysztof Wodiczko*, 148.
124 Wodiczko, *Critical Vehicles*, 47.

125 Farman, *Mobile Interface Theory*, 45.
126 Giannachi and Kaye, *Performing Presence*.
127 Farman, *Mobile Interface Theory*.
128 Because of a growing interest in Augmented Reality and the popularity of Streetmuseum, several other apps for viewing photographs and images relating to location have become available. One that has been garnering considerable attention is Local Projects' Explore 9/11 app, which enables users to contribute photos on and around September 11 of the area surrounding the former World Trade Center.
129 Another artist's work which would illustrate the connection between the two is Shimon Attie, whose work is even more closely akin to Streetmuseum. In his *The Writing on the Wall* (1991-1993) project, Attie projects pre-World War II photographs of Jewish inhabitants of Berlin's former Jewish quarter, the Scheunenviertel neighborhood, onto the location in which the photograph was originally taken.
130 Farman, *Mobile Interface Theory*, 39.
131 Ibid, 43-44.
132 Butler, *Gender Trouble*, 176.
133 Refer to Chapter 2 where I apply Derrida's notion of the supplement to the exclusion/erasure of the material with the ascendency of a concept of the virtual.
134 Farman, *Mobile Interface Theory*, 53.
135 Friedberg, *Virtual Window*, 11.
136 Ibid.
137 Theodor Adorno, *Notes to Literature* (1991), 37.
138 Site-specific media art address the sociocultural conditions of site, reveal its disciplinary effects, and highlight a differentiation of the material and the virtual.
139 Janet Cardiff, Miller and Alberro, *Janet Cardiff & George Bures Miller: Works from the Goetz Collection* (2012), 27.
140 Janet Cardiff, Miller, and Christov-Bakargiev, *Janet Cardiff: A Survey of Works Including Collaborations with George Bures Miller* (2001), 33.
141 Ibid, 158.
142 Jonathan Crary, *Suspensions of Perception* (1999).
143 I elaborate in more detail on this idea in Chapter 6.
144 Crary, *Suspensions of Perception*, 36-38.
145 See, for example, the award-winning Gettysburg Driving Tour (http://www.gettysburgdrivingtour.com), which takes listeners on a driving tour of historical sites around Gettysburg, Pennsylvania.
146 Janet Cardiff and Miller, *Her Long Black Hair*.
147 Janet Cardiff, Miller, and Christov-Bakargiev, *Janet Cardiff: A Survey of Works Including Collaborations with George Bures Miller*.
148 Atom Egoyan and Janet Cardiff, "Janet Cardiff" (2002), 65.
149 Ibid, 65.
150 I consider at greater length outdoor screens in the Conclusion.
151 Over the years members of MAW have included Ali Momeni, Jenny Schmid, Andrea Steudel, Davey Steinman, Ben Moren, Daniel Dean, Aaron Marx, and many others.
152 Nicolas Bourriaud, *Relational Aesthetics* (2002), 113.
153 I draw in part on Benford and Giannachi's account of hybridization here. See Steve Benford and Gabriella Giannachi's *Performing Mixed Reality* (2011).
154 Juleanna Enright, "Film Noir Meets Bike Culture."
155 All places, of course, have social and historical meaning, just one not evident to a general public. As a testament to how marginal these spaces were, however, while participating in *The Parade* and

Second Bridge I had never been to most of the sites before.
156 Interview with Daniel Dean and Ben Moren.
157 Ibid, 27.
158 Benford and Giannachi, *Performing Mixed Reality*, 23.
159 David Bordwell, *The Classical Hollywood Cinema* (1985), 4.
160 Ibid, 7.
161 Hito Steyerl, *The Wretched of the Screen* (2012), 32.
162 At 8.2 feet to be precise - http://en.wikipedia.org/wiki/Google_Street_View
163 See Virilio, *Speed and Politics* (2006).
164 Wendy Chun, *Control and Freedom* (2006).
165 Manovich quoted in Chun. Ibid, 39.
166 David Harvey, *Spaces of Capital* (2001).
167 Helen McLure, "The Wild, Wild Web," 475.
168 Harvey, *Spaces of Capital*, 80.
169 Giovanni Arrighi, "Hegemony Unravelling," 53.
170 Harvey, *Spaces of Capital*, 319.
171 Karl Marx, *Communist Manifesto* (1998), 39.
172 Harvey, *New Imperialism*, 123-124. Harvey has written about the American frontier only indirectly. In *Spaces of Capital* Harvey suggests that the American frontier constituted a spatial fix to the excess capacities of European capitalism in an analysis of Johann Heinrich Von Thünen, a German economic geographer.
173 Harvey, *Spaces of Capital*. 237.
174 Sidney Reed et al., "DARPA Technical Accomplishments," 28. This amount is approximately $45 million in 2012 US dollars.
175 Vint Cerf, "Amount of funding from ARPA?" (2009). This amount is approximately $90 - $125 million in 2013 US dollars.
176 Amy Schatz, "Broadband Stimulus Funds Up for Grabs" (2009).
177 Google, "Google Inc. Announces First Quarter 2013 Results" (2013).
178 See the characterization of the internet as "social factory," in Schultz's "Digital Labor" and Terranova ADD.
179 See Anne Friedberg, *Window Shopping* and Oliver Grau, *Virtual Art*.
180 Friedberg, *Window Shopping*, 143.
181 Crary, *Suspensions of Perception*.
182 Ibid, 37.
183 Marc Andrejevic, *iSpy* (2007).
184 Jonathan Crary, "Spectacle, Attention, Counter-Memory" (1989), 97.
185 Simon Sadler, *The Situationist City* (1998), 93.
186 Guy Debord, "Theory of the Dérive."
187 Sadler, *The Situationist City*, 93.
188 Debord, "Theory of the Dérive."
189 Ibid.
190 Urban Screens, "Urban Screens."
191 Daniel Schiller, *How to Think about Information* (2007).
192 Manuel Castells, *The Rise of the Network Society* (1996).
193 Sherry Turkle, *Life on the Screen* (1995).

Bibliography & Illustration Credits

ARCHIVAL SOURCE

National Archive, Kew, London, UK. Abbreviated NA in the references. The four letter code followed by a series of numbers refers to the piece reference used at the Archive.

BIBLIOGRAPHY

Adorno, Theodor, and Rolf Tiedemann. *Notes to Literature*. New York: Columbia University Press, 1991. Print.

Althusser, Louis, and Ben Brewster. *For Marx*. London; New York: Verso, 2005. Print.

Andrejevic, Mark. *ISpy: Surveillance and Power in the Interactive Era*. Lawrence, Kan.: University Press of Kansas, 2007. Print.

Apter, Emily S. *Continental Drift: From National Characters to Virtual Subjects*. Chicago, Ill.: University of Chicago Press, 1999. Print.

Armament Division Royal Aircraft Establishment at Farnborough. *Notes on Reflector Gunsight Installations using Separate Projector and Reflector Units*. NA:AVIA 15/1195 Vol., November 5, 1942. Print.

Arrighi, Giovanni. "Hegemony Unravelling." *New Left Review*. 32 (2005): 23-82. Print.

Azuma, Ronald, et al. "Recent Advances in Augmented Reality." *IEEE Computer Graphics and Applications*. 21.6 (2001): 34. Print.

Azuma, Ronald. "A Survey of Augmented Reality." *Presence: Teleoperators and Virtual Environments* 6.4 (1997): 355. Print.

Baudrillard, Jean. *Perfect Crime*. London: Verso, 2008. Print.

Benford, Steve, and Gabriella Giannachi. *Performing Mixed Reality*. Cambridge, Mass.: MIT Press, 2011. Print.

Bergson, Henri, Nancy Margaret Paul, and William Scott Palmer. *Matter and Memory*. London: G. Allen & Unwin, 1911. Print.

Bolter, Jay, and Richard A. Grusin. *Remediation: Understanding New Media*. Cambridge, Mass.: MIT Press, 1999. Print.

Bolter, Jay, et al. "New Media and the Permanent Crisis of Aura." *Convergence: The International Journal of Research into New Media Technologies* 12.1 (2006): 21-39. Print.

Bordwell, David, Janet Staiger, and Kristin Thompson. *The Classical Hollywood Cinema: Film Style & Mode of Production to 1960*. New York: Columbia University Press, 1985. Print.

Bourriaud, Nicolas. *Relational Aesthetics*. Dijon, France: Les Presses du Réel, 2002. Print.

Butler, Judith, Ernesto Laclau, and Slavoj Žižek. *Contingency, Hegemony, Universality: Contemporary Dialogues on the Left*. London: Verso, 2000. Print.

Butler, Judith. *Gender Trouble: Feminism and the Subversion of Identity*. New York: Routledge, 2006. Print.

Cameron, Chris. "Military-Grade Augmented Reality could Redefine Modern Warfare." *New York Times* June 12 2010. Print.

Cardiff, Janet, George Bures Miller, and Carolyn Christov-Bakargiev. P.S.1 Contemporary Art Center. "Janet Cardiff: A Survey of Works Including Collaborations with George Bures Miller" New York, NY: 2001. Print.

Cardiff, Janet, George Bures Miller, and Alexander Alberro. *Janet Cardiff & George Bures Miller: Works from the Goetz Collection*. München: Stiftung Haus der Kunst, 2012. Print.

Cardiff, Janet, and George Bures Miller. "Her Long Black Hair." Web. <http://www.cardiffmiller.com/artworks/walks/longhair.html>.

Castells, Manuel. *The Rise of the Network Society*. Malden, Mass.: Blackwell Publishers, 1996. Print.

Castronova, Edward. *Exodus to the Virtual World: How Online Fun is Changing Reality*. New York: Palgrave Macmillan, 2007. Print.

Cavell, Stanley. *The World Viewed; Reflections on the Ontology of Film*. New York: Viking Press, 1971. Print.

Cerf, Vint. *Amount of Funding from ARPA?* 2009. <http://mailman.postel.org/pipermail/internet-history/2009-December/001117.html>. Print.

Chun, Wendy Hui Kyong. *Control and Freedom: Power and Paranoia in the Age of Fiber Optics*. Cambridge, Mass.: MIT Press, 2006. Print.

Clarke, R. Wallace. *British Aircraft Armament 2 Vol*. London: Patrick Stephens Ltd., 1994. Print.

—. "Drawing a Bead: Part 3." *Aeroplane Monthly* April 1983 1983 Print.

Commerce Business Daily. "Sources Sought Market Survey for Land Warrior System C41 Components." Commerce Business Daily December 1 1993. Print.

Crary, Jonathan. "Spectacle, Attention, Counter-Memory." October 50 (1989): 97-107. Print.

Crary, Jonathan. *Suspensions of Perception: Attention, Spectacle, and Modern Culture*. Cambridge, Mass.: MIT Press, 1999. Print.

de Souza e Silva, Adriana. "Cyber to Hybrid: Mobile Technologies as Interfaces of Hybrid Spaces." *Space and Culture* 9.3 (2006): 261-278. Print.

Debord, Guy. "Theory of the Dérive". Ed. http://www.cddc.vt.edu/sionline/si/theory.html. Web.

Defense Advanced Research Projects Agency (DARPA). "Urban Leader Tactical Response, Awareness & Visualization (ULTRA-Vis)." June 16, 2008 2008. Web. <http://www.darpa.mil/i2o/programs/uvis/uvis.asp>.

Derrida, Jacques. *Dissemination*. Chicago: University of Chicago Press, 1981. Print.

Derrida, Jacques. *Of Grammatology*. Baltimore: Johns Hopkins University Press, 1976. Print.

—. *Specters of Marx: The State of the Debt, the Work of Mourning, and the New International*. New York: Routledge, 1994. Print.

Deutsche, Rosalyn. "Architecture of the Evicted." *Exit Art, New York City Tableaux: Tompkins Square*. Exit Art, 1990. Print.

Edwards, Paul. *The Closed World: Computers and the Politics of Discourse in Cold War America*. Cambridge, Mass.: MIT Press, 1997. Print.

Egoyan, Atom, Cardiff, Janet. "Janet Cardiff." *BOMB*.79 (2002): 60-7. Print.

Eitner, Lorenz. *Neoclassicism and Romanticism, 1750-1850; Sources and Documents*. Englewood Cliffs, N.J.: Prentice-Hall, 1970. Print.

Enright, Juleanna. "Film Noir Meets Bike Culture." *l'etoile magazine*. June 7 2012. Print.

Farman, Jason. *Mobile Interface Theory: Embodied Space and Locative Media*. New York: Routledge, 2012. Print.

Farren WS. "Letter to the Ministry of Aircraft Production." NA: AVIA 15/1195. 9 February 1941. Print.

Foster, Hal. *Discussions in Contemporary Culture. Number One*. Seattle: Bay Press, 1987. Print.

Freud, Sigmund, Josef Breuer, and James Strachey. *The Standard Edition of the Complete Psychological Works of Sigmund Freud*. London: Hogarth, 1955. Print.

Friedberg, Anne. *Virtual Window*. Cambridge, Mass.: MIT Press, 2006. Print.

—. *Window Shopping: Cinema and the Postmodern*. Berkeley, Cal.: University of California Press, 1994. Print.

General Dynamics. "Land Warrior." 2010. Web. <http://www.gdc4s.com/documents/landwarrior_web010.pdf>.

Giannachi, Gabriella, and Nick Kaye. *Performing Presence: Between the Live and the Simulated*. Manchester, UK: Manchester University Press, 2011. Print.

Google. "Google Inc. Announces First Quarter 2013 Results." April 18 2013. Web. <http://investor.google.com/earnings/2013/Q1_google_earnings.html>.

Grau, Oliver. *Virtual Art: From Illusion to Immersion*. Cambridge, Mass.: MIT Press, 2004. Print.

Gray, Chris Hables. *Postmodern War: The New Politics of Conflict*. New York: Guilford Press, 1997. Print.

Hansen, Mark. *Bodies in Code*. New York: Routledge, 2006. Print.

Haraway, Donna. *Simians, Cyborgs, and Women: The Reinvention of Nature*. New York: Routledge, 1991. Print.

Harvey, David. *The New Imperialism*. Oxford; New York: Oxford University Press, 2005. Print.

—. *Spaces of Capital: Towards a Critical Geography*. New York: Routledge, 2001. Print.

Hayles, N. Katherine. *Electronic Literature: New Horizons for the Literary*. Notre Dame, Ind.: University of Notre Dame, 2010. Print.

—. *How we Became Posthuman: Virtual Bodies in Cybernetics, Literature, and Informatics*. Chicago, Ill.: University of Chicago Press, 1999. Print.

Hayles, N. Katherine. "The Materiality of Informatics." *Configurations* 1.1 (1993): 147. Print.

Herczka, Mateusz, and Pär Frid. *Reverse Avatar*. 2010.

Instrument & Photography Department Royal Aircraft Establishment at Farnborough. *Bullet Proof Screen as Gunsight Reflector*. NA:AVIA 15/1195 Vol. May 1943. Print.

Jones, Caroline, and Bill Arning. *Sensorium: Embodied Experience, Technology, and Contemporary Art*. Cambridge, Mass.: MIT Press, 2006. Print.

Jurgenson, Nathan. "Digital Dualism versus Augmented Reality." *The Society Pages*. February 24 2011. Web. <http://thesocietypages.org/cyborgology/2011/02/24/digital-dualism-versus-augmented-reality/>.

Kim, John. "The Origin of the See-through Graphical Interface: World War II Aircraft Gunsights and the Status of the Material in Early Computer Interface Design." *Convergence: The International Journal of Research into New Media Technologies* (May 2015) Print.

Kwon, Miwon. *One Place After Another: Site-Specific Art and Locational Identity*. Cambridge, Mass.: MIT Press, 2004. Print.

Layar. "What is Layar."Web. <http://site.layar.com/download/layar/>.

Leigh-Mallory, Trafford. *Pilot's Gyro Gunsight Mark II*. NA:AIR 20/1739 Vol. October 27 1943. Print.

Lenoir, Timothy. "Fashioning the Military Entertainment Complex." *Correspondence: An International Review of Culture and Society* 10. Winter/Spring (2002-2003): 14-16. Print.

Lévy, Pierre. *Becoming Virtual: Reality in the Digital Age*. New York: Plenum Trade, 1998. Print.

Lister, Martin, et al. *New Media: A Critical Introduction*. London: Routledge, 2009. Print.

Livingston, Mark, et al. "An Augmented Reality System for Military Operations in Urban Terrain". December 2-5, 2003, Orlando, Flor., 2003. 868–875. Print.

Lockyer, CCW. *Letter to Ministry of Aircraft Production*. NA:AVIA 15/1195 Vol. March 30 1943. Print.

Mann, Steve. "Wearable Computing as Means for Personal Empowerment." *International Conference on Wearable Computing*. May 12-13, 1998. Print.

Manovich, Lev. *The Language of New Media*. Cambridge, Mass: MIT Press, 2002. Print.

Martin, Reinhold. "Occupy: The Day After." *Design Observer*. 12/08/2011 2011.Web. <http://places.designobserver.com/feature/occupy-the-day-after/31698/>.

—. "Occupy: What Architecture Can Do." *Design Observer*. 11/07/2011 2011. Web. <http://places.designobserver.com/feature/occupy-what-architecture-can-do/31128/>.

Marx, Karl, Friedrich Engels, and E.J. Hobsbawm. *The Communist Manifesto: A Modern Edition*. London; New York: Verso, 1998. Print.

McLuhan, Marshall. *Understanding Media; the Extensions of Man,*. New York: McGraw-Hill, 1964. Print.

McLure, Helen. "The Wild, Wild Web: The Mythic American West and the Electronic Frontier." *The Western Historical Quarterly*. 31.4 (2000): 457. Print.

Milgram, Paul, and Fumio Kishino. "Taxonomy of Mixed Reality Visual Displays." *IEICE Transactions on Information and Systems* E77-D.12 (1994): 1321-1329. Print.

Mindell, David. *Between Human and Machine: Feedback, Control, and Computing before Cybernetics*. Baltimore: Johns Hopkins Press, 2004. Print.

Moy, Timothy. *War Machines: Transforming Technologies in the U.S. Military, 1920-1940*. College Station, Texas: TAMU Press, 2001. Print.

Munster, Anna. *Materializing New Media: Embodiment in Information Aesthetics*. Hanover, N.H.: Dartmouth College Press, 2006. Print.

Naish, JM. *The Head-Up Steering Display for the T.S.R. II Aircraft*. NA:AVIA 6/19907 Vol. , December 1959. Print.

—. *A Study of O.R. 3044 for a Navigation, Bombing Reconnaissance and Flight Control System for Strike Reconnaissance Aircraft*. NA:DSIR 23/26510 Vol. January 1958. Print.

Oxford University Press. "Oxford English dictionary." 2000. Web.

Poster, Mark. *What's the Matter with the Internet*. Minneapolis, MN: University of Minnesota Press, 2001. Print.

Price, Alfred. *Late Marque Spitfire Aces: 1942-45*. Oxford, UK: Osprey Publishing, 1998. Print.

Reed, Sidney G., Richard H. Van Atta, and Seymour J. Deitchman. *DARPA Technical Accomplishments: An Historical Review of Selected DARPA Projects*. Alexandria, Va.: Institute for Defense Analyses, 1990. Print.

Rheingold, Howard. *Virtual Reality*. New York: Summit Books, 1991. Print.

Royal Aircraft Establishment at Farnborough. *Gyro Gunsight Adoption by U.S.A.: Development, and Firing Trials in America*. NA:AVIA 15/1206 Vol. December 1941. Print.

Royle, Nicholas. *The Uncanny*. New York: Routledge, 2003. Print.

Sadler, Simon. *The Situationist City*. Cambridge Mass.: MIT Press, 1998. Print.

Schatz, Amy. "Broadband Stimulus Funds Up for Grabs." *Wall Street Journal*. April 15 2009. Print.

Schiller, Dan. *How to Think about Information*. Urbana: Ill.: University of Illinois Press, 2007. Print.

Scholz, Trebor. *Digital Labor: The Internet as Playground and Factory*. New York: Routledge, 2013. Print.

Schopenhauer, Arthur, et al. *The World as Will and Representation*. Cambridge; New York: Cambridge University Press, 2010. Print.

Science News Letter. "Gunsight for Planes." *Science News Letter*. August 26 1944: 130. Print.

Shields, Rob. *The Virtual*. London; New York: Routledge, 2003. Print.

Silverman, Kaja. *The Acoustic Mirror: The Female Voice in Psychoanalysis and Cinema*. Bloomington, Ill.: Indiana University Press, 1988. Print.

Sinclair, Archibald. *Letter from Archibald Sinclair, Air Ministry, to J.J. Llewellin, Ministry of Aircraft Production*. NA:AIR 20/1739 Vol. September 10 1942. Print.

Skoller, Jeffrey. *Shadows, Specters, Shards: Making History in Avant-Garde Film*. Minneapolis, MN: University of Minnesota Press, 2005. Print.

Smith, Neil. "Contours of a Spatialized Politics: Homeless Vehicles and the Production of Geographical Scale." *Social Text*. 33 (1992): 55-81. Print.

Steyerl, Hito. *The Wretched of the Screen*. Berlin: Sternberg Press, 2012. Print.

Sutherland, Ivan. "A Head-Mounted Three-Dimensional Display". *Proceedings of American Federation of Information Processing Societies*. 1968. Print.

Tan, Michelle. "Pioneering Stryker Unit Preps for Afghanistan." *Army Times*. April 21 2009. Print.

Taylor, Joy. "Nightmare Japan: Contemporary Japanese Horror Cinema by Jay McRoy." *Journal of Popular Culture* 43.1. 2010. Print.

Turkle, Sherry. *Life on the Screen: Identity in the Age of the Internet*. New York: Simon & Schuster, 1995. Print.

Urban Screens. "Urban Screens. "Web. <http://urbanscreens.org/>.

Virilio, Paul. *Speed and Politics: An Essay on Dromology*. Los Angeles:

Semiotext(e), 2006. Print.

Virilio, Paul. *War and Cinema: The Logistics of Perception*. London: Verso, 1989. Print.

Ward, Mark, et al. "A Demonstrated Optical Tracker with Scalable Work Area for Head-Mounted Display Systems". *Proceedings of the 1992 Symposium on Interactive 3D Graphics*. 1992. 43-52. Print.

Weintraub, Daniel, and Michael Ensing. *Human Factors Issues in Head-Up Display Design: The Book of HUD*. Dayton, Ohio: University of Dayton Research Institute, 1992. Print.

Whatmore, Sarah. "Materialist Returns: Practicing Cultural Geography in and for a More-than-Human World." *Cultural Geographies* 13.4 (2006): 600-9. Print.

Williams, Raymond. *Television : Technology and Cultural Form*. Hanover, N.H.: Wesleyan University Press, 1992. Print.

Wing Commander (name unknown). *Letter from the Wing Commander of the Air Fighting Development Unit, Royal Air Force to the Ministry of Aircraft Production*. NA:AVIA 15/1195 Vol. March 7 1943. Print.

Wodiczko, Krzysztof, and Duncan McCorquodale. *Krzysztof Wodiczko*. London: Black Dog Pub Ltd, 2011. Print.

Wodiczko, Krzysztof. *Critical Vehicles: Writings, Projects, Interviews*. Cambridge, Mass.: MIT Press, 1999. Print.

Wodiczko, Krzysztof. "Projections." *Perspecta* 26 (1990): 273-87. Print.

Elephant. Dir. Gus Van Sant. HBO Video, 2003.

ILLUSTRATION CREDITS

Mateusz Herczka and Pär Frid, "Reverse Avatar," 2010. Used with permission.

Mateusz Herczka and Pär Frid, "Reverse Avatar" alternative view, 2010. Used with permission.

"Land Warrior" ©PEOSoldier. Licensed under CC BY 2.0. https://www.flickr.com/photos/peosoldier/3882312870/

Cockpit of a spitfire airplane. Photography by Rottweiler in the public domain. http://commons.wikimedia.org/wiki/File:Spitfire_cockpit.jpg

R. Wallace Clarke, British Aircraft Armament (1994), 171. Diagram adapted by author.

Google Glass™. Google and the Google logo are registered trademarks of Google Inc., used with permission.

"Arco de la Victoria, Madrid, 1991" © Krzysztof Wodiczko. Courtesy Galerie Lelong, New York.

"Tilted Arc, 1981" © G. Lutz. Used with permission.

Janet Cardiff and George Bures Miller, "Paradise Institute," 2001. Used with permission.

Janet Cardiff and George Bures Miller, alternate view of "Paradise Institute," 2001. Used with permission.

Janet Cardiff and George Bures Miller, "Her Long Black Hair," 2004. Used with permission.

Graffiti Research Lab's Mobile Broadcast Unit. Photography by Graffiti Research Lab, Evan Roth, and James Powderly in the public domain. https://www.flickr.com/photos/urban_data/452740396/

Minneapolis Art on Wheels' iteration of the Mobile Broadcast Unit. Used with permission.

A screenshot of Livedraw. Software by Minneapolis Art on Wheels in the public domain.

Minneapolis Art on Wheels, "Seaworthy," 2010. Used with permission.

Mobile Experiential Cinema, "Second Bridge is Wider, but not Wide Enough," 2012. Illustration by Trevor Burks. Used with permission.

Mobile Experiential Cinema, "The Parade," 2013. Photography by Ben Moren. Used with permission.

John Kim, "Airplane Tree Minaret," 2003.

John Kim, "Military Bay," 2003.

Index

absence 6, 8-9, 24
abstraction 5
Adorno, Theodor 93
Advanced Research Projects Agency (ARPA) 134
aesthetic, aestheticism 15, 17, 82, 92, 115, 124, 136, 152
Airplane-Tree-Minaret 139, 141
Afghanistan 49
algorithm 46-47, 130
alienation 103
alternative, alternate 70, 85, 104, 115, 141, 147, 149
Althusser, Louis 80
analog 47, 57
Andrejevic, Mark 157
Android (Operating System) 63
angle of deflection firing 56-58
appropriation 25, 148
Apter, Emily 31-32
arcade 136
architecture 70, 77, 79-80
Arco de la Victoria 75, 77, 94
Arrighi, Giovanni 133
artificial 8-9
Atlantic Ocean 128
attention, attentive, attentiveness 30, 54, 88, 99-100, 123, 129, 136-137, 140-141, 147; Attention Deficit Disorder 100
audio walk 101, 105
augment, augmented, augmentation 2, 5, 10-11, 38, 50, 63
Augmented Reality (AR) 5, 12, 14-15, 21, 30, 44, 45, 48, 52, 67, 82-83, 123, 124, 142
avant-garde 77, 138
awareness, critical 12-13, 17, 22, 89, 105, 124, 141, 142, 147, 150; situational 52; total information 64
Azuma, Ronald 45-46

background (interface) 44, 76
ballistics 62
Barlow, John Perry 132
Baudrillard, Jean 25-28, 85
Benford, Steve 113-114
Bentham, Jeremy 79
Bergson, Henri 34
binary (numbers) 47
binaural 103
Blast Theory 113
body, bodies 21, 31-37, 50, 103
Bolter, Jay 9-10
bombsight 58; Norden bombsight 58
Bordwell, David 114
boundary (between virtual and real) 21, 98
bourgeois, bourgeoisie 133-134, 138
Bourriaud, Nicolas 109
bulletproof 60, 62
Bures Miller, George 16, 92, 97-105, 116, 148-149
Butler, Judith 36

camera 2, 51, 52,
camera obscura 28, 152
Canadian Pavilion (Venice Biennale 2001) 98
capital, capitalist, capitalism 26, 27, 35, 115, 131-135, 138, 139, 141, 148, 149, 151; spatio-temporal fix 131-135
Cardiff, Janet 16, 92, 97-105, 116, 148-149
Castells, Manuel 148
cathode ray tube 54
Cavell, Stanley 20, 29
cellphone 33
Cerf, Vint 134

Chardin, Jean-Baptiste-Siméon 6-7
China 134
Chun, Wendy 130
cinema 14, 20, 58, 88, 102, 112, 123, 136, 152; cinematic 29, 51, 77, 99, 110, 111, 136; "physical cinema" 102
cinematography 21, 114
cinéorama 136
Clarke, R. Wallace 58, 67
cockpit (airplane) 53-55, 60-62, 66
colonization 86, 131
combat 50-53, 56, 59, 64
commodity 26, 27
communication 5, 8-9, 31, 33-34, 50-51, 135, 137; internet protocols 130, 134
compression (software) 128
computer, computers, computing 2, 3, 4, 14, 21, 22, 25, 28, 31, 35, 45, 46, 48, 51, 52, 53, 99, 102, 113, 123, 125, 130, 132, 137; "computer unit" 57; desktop 16, 70
constructionism, cultural 35-36
constructivism, cultural 33
contingency 22
Crary, Jonathan 99-100, 136-137
critical, critique 12-13, 15-17, 70, 74, 75, 78, 79, 81-82, 90-93, 96, 115, 124, 126, 127, 138-142, 147, 150, 152; critical practice 12-14, 16-17, 22, 64-65, 70, 74, 90, 92-93, 138-139, 141-142, 147-152
culture 22-27, 37, 85,

91, 138, 147
cybernetic 5, 50
cyberspace 25, 130
cyborg 33, 50

Dean, Daniel 92, 110-115, 158
Debord, Guy 80, 139-140
"Declaration of Independence of Cyberspace" 132
deconstruction 10, 22, 35
Defense Advanced Research Projects Agency (DARPA) 134
deflection (lead or angle firing) 56-58
dematerialize 27, 85
dérive 111, 123, 138-140
Derrida, Jacques 7-10, 24-28, 37, 85
destabilize 24-26
détournement 138, 140
Deutsche, Rosalyn 74-75
diacritical 8
dialectic 7-8
Diderot, Denis 6-7
Dietz, Steve 106
différance 8
digital 31, 33, 46, 47, 49, 51, 82, 83, 85, 100, 107, 111, 128, 135, 141; communications 31; computer 4; economy 129, 137, 141; labor 137; media 13, 22, 148
disappearance 11, 13, 37, 76, 85-86
discipline, disciplinary 13, 15, 70, 74, 78-82, 99, 129, 136-141, 150
disembody, disembodiment 31-33, 46, 85, 129
disorient, disorientation 8, 21, 100, 101, 105, 112, 122, 125

display (see Head Down Display, Head Mounted Display, Head-up Display)
dissimulate 9
Do It Yourself (DIY) 105
docile, docility 122-123, 129, 131, 135, 137-138, 140
dogfight 57
dream 122, 138
drone 127
drug 2, 8, 50
dualism, dualistic 46-47, 64

economy, economic 17, 23, 25-27, 80, 123, 129, 132-135, 137-138, 141; attention economy 100
electromagnet 57
Elephant (movie) 21
emancipation 33,
embedded 23, 40, 91, 128, 151
embody, embodied, embodiment 13, 22, 31, 33-36, 65
encode 47
encounter 13, 16, 17, 65, 70, 74, 89-91, 93, 105, 109, 115, 124, 128, 129, 140, 141, 150, 152
enframing 28
enhance, enhancement 2-4, 11, 14, 21, 33-34, 41, 44, 45, 49-51, 53-55, 59, 63, 85, 88, 91, 147
environment 2, 4, 15, 20, 26, 28-29, 35-36, 39, 44, 46-49, 52-53, 60, 61, 63, 70, 74, 79-91, 97, 99, 102-104, 113-114, 127, 128, 146, 148, 151-152
ephemeral 82
epistemic 27
epistemological 11,

38, 40
essentialize, essentializing 31, 33, 35
excess 133
exclusion 11, 13-14, 22-24, 27, 30-31, 35, 37, 85, 147
extension 2-3, 34, 59, 97, 147
eye 6-7, 14, 51, 54, 122, 129
eyeglasses 38
eyepiece 52, 58

facade 71, 112
factory (digital) 135
fantasy 50, 138
farming 150-151
Farman, Jason 15, 30, 47-48, 83-84, 86
fiber optic 135
film 2, 10, 34, 77, 112, 114-115; filmmaking 77, 110
financial, finance 27, 134
flâneur 136
flesh 26
flows (media) 148
food 150-151
foreground 10, 16, 27, 37, 38, 74, 89-92, 98, 128
Foucault, Michel 79, 137
Franco, Francisco 76
freedom 123
Freud, Sigmund 71-72
Frid, Pär 20-21, 42
Friedberg, Anne 3, 23, 28-29, 73, 87, 135-137
frontier 131-132

Galerie Lelong 94
gaze 30, 123
gender 31, 33, 36
gentrification 75
geography 133; geographical location 30, 48, 51, 52, 83, 84, 127, 131
ghost, ghostly 24-25, 72-73; see also

specter
Giannachi, Gabriella 15, 83, 113, 114
Glow Festival 146
goggles (video) 21
Google 63, 130, 135; Google Glass 44, 63, 67, 149; Google Street View 111, 129
Global Positioning Satellite (GPS) 51, 113, 125
Graffiti Research Labs (GRL) 106, 117
Graphical User Interface (GUI) 28-29, 53; see also See-Through Graphical Interface
graticule 57-58, 60, 67
Grau, Oliver 29, 135
Gray, Chris Hables 50
Greeley, Horace 131
Grusin, Richard 9-10
guerilla 107, 108
Gulf War 75, 126
gun 51, 58, 75
gunnery 56
gunsight 14, 44, 49, 51, 54-63; gyro gunsight 14, 53-62, 66
gyroscope 57

habitus 142
hacker 132
Hansen, Mark 33-35
happenings 106
Haraway, Donna 33
hardware 44, 49, 130, 134-135
Harvey, David 131-134
haunted 71-73
hauntology 23
Hayles, Katherine 32
Head Down Display (HDD) 54
Head Mounted Display (HMD) 51-53, 61-63, 67, 102, 124
Head-up Display (HUD) 4, 14-15, 44, 49, 53-54, 62

Heathrow Airport 122
Hebdige, Dick 71, 75, 80
hegemony 80
Hennepin Bridge 111
Herbicide 151
Herczka, Mateusz 20-21, 42
Hollywood 99, 113-114
homeless, homelessness 72, 75
hybrid, hybridizes, hybridization 15, 50, 83, 86, 96, 110, 113-115
hyperactivity 100
Hyperlapse 129-130

ideology 50, 123
illusion 88
immaterial 3, 17, 20, 24-27, 30, 31, 34, 40, 46, 73, 74, 82, 87, 89, 90
immediate, immediacy 2, 3, 4, 6-10, 13, 15, 20, 38-41, 52, 147, 152
immersion, immersed, immersive 16, 25, 28-29, 99, 101, 102, 104, 106, 147
immobile, immobility 48, 70, 84, 123, 133
imperfect, imperfections 128, 141
impermanence 87
implant 2
indexical 30, 48, 84
informatics 13, 23, 31
information 3, 25, 28, 30-34, 44, 47-49, 51-64, 70, 84-86, 90, 114, 127, 130
interactive 10, 21, 22, 28, 29, 105, 123, 146; interaction 9, 11, 15, 20, 31, 33, 37, 38, 46-48, 52, 64, 82, 83, 86, 89, 98, 109, 141-142
interface 3, 9, 14, 21, 22, 30-31, 44-45, 48, 53, 60-63, 82, 84; see

also Graphical User Interface and See-Through Graphical Interfaces
internet 33, 86, 129, 130-132, 134-135, 137
interpellate 80
interrogative 15, 73-74, 76-78, 81, 91-92, 128

Jones, Caroline 2

Kim, John 143
Kishino, Fumio 45
Kwon, Miwon 73-74, 76-78, 81

Land Warrior System (LWS) 49-53, 66
layer 48-49, 58, 64, 107, 130
lead firing 56-58
Leigh-Mallory, Trafford 55
liminal 3, 20, 73, 87
live 15, 21, 102, 110, 112-113
Livedraw 107, 118

location 13, 16, 30, 48, 75, 78, 83-84, 96, 101, 108, 109, 111, 130, 138, 140, 150
locative 48
London 83
Luftwaffe 55
lyric poetry 93

Mann, Steve 45
Manovich, Lev 47
Mark II Gyro Gunsight 14, 53-59, 62-63
Marx, Aaron 158
Marx, Karl 132
material 3-5, 9, 11-17, 21-40, 44-50, 53, 58-61, 63-65, 73, 80-93, 96-98, 102, 105, 109, 110, 113, 115, 124, 126-129, 140-142, 146-152; materialism 19;

materialist 27, 39-40, 89; materiality 31-32, 34-37, 65, 70, 73, 74, 78, 82, 87, 89-91, 98, 111, 115, 128, 141, 147, 150, 151
Minneapolis Art on Wheels (MAW) 16, 105-109
MaxMSP 107
McLuhan, Marshall 2
McLure, Helen 131, 132
mediated 2, 7-10, 13, 28, 45, 47, 85, 97, 102, 113, 127, 148; mediation 9-10, 17, 37, 151; mediare 6
Merleau-Ponty, Maurice 34
Messerschmitt (German aircraft) 56
metaphor 8, 26, 87, 132-133
Milgram, Paul 45
military 14-15, 44, 49-55, 59, 62-64, 70, 125-127, 133, 140, 149, 152; militarism 75, 76, 78, 80; militarization 75
Military Bay 125-129, 139-141, 143
Miller, George Bures 16, 92, 97-105, 116, 148-149
miniaturization 53
Minneapolis 106, 108, 111, 146
Minnesota (see Minneapolis)
Mississippi River 112
mixed reality 30, 83, 113-114
Mobile Broadcasting Unit (MBU) 106-108, 110
Mobile Experiential Cinema (MEC) 110-115
mobile, mobility 21, 30, 45, 47-52, 63, 70, 82-84, 86, 105-107, 109, 111, 113, 123, 124, 128, 129, 131, 139, 148; virtual mobility 129, 135-138
modern 3, 4, 5, 9, 12, 30, 79, 99-100, 129, 135-138, 141; modernist 32
Momeni, Ali 106-108
montage 77
monument 70, 71, 76-80, 88
Moren, Ben 92, 110-115, 158
movement 70-71, 76-80, 88
Munster, Anna 32-33
myth 33, 79, 81

Naish, J.M. 54, 62
narrative 92, 99, 100, 101, 104-105, 110-115
nature 6
navigation 53, 139
network (information) 52, 113
non-place 101, 114
Norden bombsight 58
Northern Spark 106, 110, 146
nuit blanche 106, 146
numerical (representation) 47-48

Obama, Barack 49
ontological 10
opaque (screen) 21, 64, 82
optical 60, 67
organic 151
organism 50
organ (sensory) 2, 5, 11
outdoor 4, 17, 70, 84, 96, 97, 105-107, 109, 115, 146
outside 8, 29, 35-37, 98-101, 104, 105, 123
overaccumulation 131-133
overlay 41, 77, 84, 88, 91, 124

painting 5-7, 28, 83, 86, 152

panopticon 79
Paradise Institute 98-101, 116
participation, participatory 98, 102, 104, 107-109, 139
peephole 58-59, 61, 63
perceptual 9, 28-30, 31, 34-35, 48, 87, 147
performance 55, 71, 92, 107, 110, 113-114
performative (act) 36
pesticide 151
Phaedrus 7
Pharmakon 8
pilot 53-63, 67
place 39, 72-73, 77, 80, 88, 97, 101, 103-104, 127, 130, 138-139, 148; "non-place" 101, 114
Plato 7-8
playground (factory as) 135
Poster, Mark 25
posthuman 32
postidentitarian 32
postmodern 26, 50; postmodernism, postmodernist 25
power 17, 25, 74, 76, 79-82, 127, 140; labor 132, 135
practice (critical) 12-14, 16-17, 22, 64-65, 70, 74, 90, 92-93, 138-139, 141-142, 147-152
presence 6, 8, 13, 24, 33, 38, 83, 110, 114, 141-142
privatization (mobile) 148-149
proscenium 88
prosthetic, prostheticized 25, 33, 50, 147
protest 83, 92, 147, 151
protocol (communication) 130, 134
psychogeography 80, 111
psychology, psychological 61, 100, 114, 123, 140
public art 70, 73, 76, 78

queer 33

race, racial 31-33
radical 138
reality 26-27, 34-35, 102, 104, 123
recession 134
reflector (screen) 57-63, 67
remediation 9-10
remote 21, 51, 127
representation 20, 87, 130, 151; see also numerical representation
repressed, repression 71, 73, 81, 138
resistant, resistance 17, 35, 115, 138, 149
retina 54
Reverse Avatar 20-21, 42
Rheingold, Howard 132
Royal Aircraft Establishment at Farnborough 54, 60, 62
rupture 13, 16, 17, 90, 93, 124, 128, 142, 150

San Francisco Bay 125-126, 140
Santa Monica 146
schemata 114
Schiller, Dan 148
schizophrenia 100
Schmid, Jenny 158
Schopenhauer, Arthur 40, 89
screen 2, 20-21, 29-30, 44, 53, 55, 57-64, 67, 70, 82, 89, 100, 109, 110, 114-115, 122-123, 128, 136-137, 140-141, 146-148
Seaworthy 109, 118
sedentary, sedentariness, sedentarization 123, 129, 137-138, 141
See-Through Graphical Interface (STGI) 3-4, 12, 14-16, 41, 43-44, 49, 55, 60-64, 70, 124, 128, 140-141, 149, 152
semiotic 104
sense, sensory (see organs)
sensorium 2
Serra, Richard 76-78, 94
signification 40, 89
simulacrum 24-25
simulate, simulated, simulation 2, 15, 97, 103, 113
site (see site-specific and site responsive)
site-specific, site-specificity 4, 12, 14-17, 69-74, 76, 78, 81, 83, 85, 88, 90, 92, 96-98, 101, 105-106, 110-112, 115, 124-125, 129, 138, 141-142, 146, 149, 150
site responsive 16, 73-75, 95-96, 105-106, 110, 112, 114-115
Situationism 111, 138-139
situational awareness 52
Sixthsense 48
SJ01 Biennial 106
slaughterhouse 122
smart bomb 127
smartphone 48, 86, 148
software 44, 46, 51, 63, 105, 107, 118, 130, 134, 135
soil 151
soldier 49-53, 63-64, 66
Sony Walkman 103, 148-149
space 46, 79, 80, 88, 92, 114, 127, 130-134, 138, 140; material or physical 21, 28, 48, 83-84, 127, 130; public 17, 77, 104; urban 17; virtual 2-3, 20, 25, 28-29, 41, 44, 83, 136, 147
spatialization 129-131, 135, 137
spatio-temporal fix 131-135
spectacle 17, 136, 139-140

spectator, spectatorship 88, 99, 123, 136-138, 141
specter, spectral 24-25, 72, 89
speed 56-57, 59, 129-130
Spitfire (airplane) 55-56, 60, 66
Steadicam 21
Steinmann, Davey 158
Steudel, Andrea 158
Steyerl, Hito 128
Streetmuseum 83, 86
subjectivity 32, 35-36
sublimate, sublimation 138
superimpose 52, 58, 61, 83
supplementarity 22-27, 37
surface 60-61, 75, 82, 87, 89, 141
surplus 132-135
surroundings (physical or material) 3-4, 12, 15, 20-22, 30, 38, 40-41, 44, 46-49, 52-53, 58-59, 64, 70, 82-85, 96-97, 100, 105, 124, 140-141, 147-148, 152
surveillance 25, 52, 130, 139
synecdoche 33
synthetic 25, 45

target, targeting 51, 54-63
TCP/IP 130
technics 11, 34
technology 3, 9, 11-16, 20-23, 25-26, 28-34, 37-38, 40, 44-51, 53-55, 58, 62-64, 70, 82-87, 90-91, 105-106, 111, 115, 122-124, 126, 127, 132, 134-136, 140, 147-149, 152; technologist 45
telepresence 45
television 2, 70, 123, 137
temporary 82, 138
territorialize 133

total information
 awareness 64
transparent, transparency 9, 53, 58-60, 63
traumatic 81
trompe l'eoil 6
Turkle, Sherry 148

ubiquitous (computing) 113
unaugmented 2
uncanny 71, 73, 88
unconscious 76, 80, 103, 115
unheimlich 71, 72, 80, 81
United States of America 126, 127, 131
unmediated 6, 7, 9, 52
unprogrammed 111-112; see also non-places
Uptown (Minneapolis neighborhood) 108
urban 17, 71, 79-80, 108, 110, 146, 150; urban screens 146

van Sant, Gus 21
Venice Biennale 98
video 15, 21, 47, 51, 52, 71, 82, 86, 105-108, 111-112, 127, 146; walks 97-98, 100-102; games 10, 28, 29, 137
Virilio, Paul 14, 51, 58
virtual 3-4, 11-17, 20-41, 44-48, 63-65, 70, 82-93, 96, 98, 102, 105, 110, 113-115, 124, 127-131, 135-137, 141-142, 146-152; virtuality 34-35, 88, 91, 110, 123; virtualization 91, 126; virtual spaces 2-3, 20, 25, 28-29, 41, 44, 83, 136, 147
Virtual Reality 25-29, 35, 45, 102-103, 148
visuality 44, 46, 51
visualization 53, 125, 130

Walker Art Center 106

Walkman (Sony) 103, 148-149
wander, wanderings 136, 138-140
war 50, 62, 75-76, 126
wearable 53, 125
Williams, Raymond 5, 148
window 2, 21, 28-29, 41, 61, 82, 122; virtual window 3, 44
windscreen 60-62
World War I 56
World War II 14, 44, 49, 53-54, 56, 62
Wodiczko, Krzysztof 14-15, 69-93

Žižek, Slavoj 35-36

COLOPHON

An earlier version of Chapter 2 "A New Materialism for New Media Studies" was published in *communication+1* as "Whither the Material in New Media Studies" (Fall 2013). A version of Chapter 3 "The Origin of the See-Through Graphical Interface" was published in *Convergence* (May 2015). A version of Chapter 4 "Encounters with the Material: Krzysztof Wodiczko and Site-Specific Media Art" was published in *Discourse* (Spring 2016). Portions of Chapter 6 "Rupture of the Virtual" were presented at the "Critical Studies in New Media" conference at Stanford University in 2004; and as a multimedia performance at the Sarai Institute in New Delhi, India during a residency there in 2009.

COPYRIGHT INFORMATION

This book is licensed under the Creative Commons Attribution-Share Alike 3.0 license. (This does not necessarily apply to the figures contained within. Please check the individual copyright designations for them.)
You are free:
to share – to copy, distribute and transmit the work
to remix – to adapt the work
Under the following conditions:
attribution – You must attribute the work in the manner specified by the author or licensor (but not in any way that suggests that they endorse you or your use of the work).

The author has made every effort to secure permission to reproduce the listed material – illustrations and photographs. We apologize for any inadvertent errors or omissions. Parties who nevertheless believe they can claim specific legal rights are invited to contact the author.

ISBN
978-0-692-71850-6

AUTHOR
John Kim

COPY EDITING
Laura Bigger
Alicia Johnson

DESIGN
Sara Fowler

ADDITIONAL FUNDING AND ASSISTANCE
DeWitt Wallace Library, Macalester College

www.ingramcontent.com/pod-product-compliance
Lightning Source LLC
Chambersburg PA
CBHW070150100426
42743CB00013B/2870